AF732250

Plantation Crops, Plunder and Power

Over the last five centuries, plantation crops have represented the best and worst of industrialized agriculture – "best" through their agronomic productivity and global commercial success, and "worst" as examples of exploitative colonialism, conflict and ill-treatment of workers. This book traces the social, political and evolutionary history of seven major plantation crops – sugarcane, banana, cotton, tea, tobacco, coffee and rubber.

It describes how all of these were domesticated in antiquity and grown by small landowners for thousands of years before European traders and colonists sought to make a profit out of them. The author relates how their development and spread were closely associated with government expansionist policies. They stimulated the exploration of far off lands, were the focus of major conflicts and led to the enslavement of both native and displaced peoples.

From the southern United States, Latin America and the Caribbean, to Asia and Africa, plantation crops turned social structures upside down leading to revolution and government change. The economies of whole countries became tied to the profits of these plantations, leading to internal power struggles to control the burgeoning wealth. Open warfare routinely broke out between the more powerful countries and factions for trade dominance. This book shows that from the early 1500s to today, at least one of the plantation crops was always at the center of world politics, and that this still continues today, for example with the development of oil palm plantations in Southeast Asia.

James F. Hancock is Professor Emeritus, Department of Horticulture, Michigan State University, USA. He is author of *Plant Evolution and the Origin of Crop Species*, now in its third edition.

Other books in the Earthscan Food and Agriculture Series

Climate Change and Agricultural Development
Improving resilience through climate smart agriculture, agroecology and conservation
Edited by Udaya Sekhar Nagothu

Forgotten Agricultural Heritage
Reconnecting food systems and sustainable development
Parviz Koohafkan and Miguel A. Altieri

International Law and Agroecological Husbandry
Building legal foundations for a new agriculture
John W. Head

Food Production and Nature Conservation
Conflicts and Solutions
Edited by Iain J. Gordon, Herbert H. T. Prins and Geoff R. Squire

Sustainable Intensification in Smallholder Agriculture
An integrated systems research approach
Edited by Ingrid Öborn, Kwesi Atta-Krah, Michael Phillips, Richard Thomas, Bernard Vanlauwe and Willemien Brooijmans

Environmental Justice and Farm Labor
Rebecca E. Berkey

Plantation Crops, Plunder and Power
Evolution and Exploitation
James F. Hancock

For further details please visit the series page on the Routledge website: http://www.routledge.com/books/series/ECEFA/

Plantation Crops, Plunder and Power

Evolution and Exploitation

James F. Hancock

Routledge
Taylor & Francis Group
LONDON AND NEW YORK

earthscan
from Routledge

First published 2017
by Routledge
2 Park Square, Milton Park, Abingdon, Oxon OX14 4RN

and by Routledge
711 Third Avenue, New York, NY 10017

Routledge is an imprint of the Taylor & Francis Group, an informa business

© 2017 James F. Hancock

The right of James F. Hancock to be identified as author of this work has been asserted by him in accordance with sections 77 and 78 of the Copyright, Designs and Patents Act 1988.

All rights reserved. No part of this book may be reprinted or reproduced or utilised in any form or by any electronic, mechanical, or other means, now known or hereafter invented, including photocopying and recording, or in any information storage or retrieval system, without permission in writing from the publishers.

Trademark notice: Product or corporate names may be trademarks or registered trademarks, and are used only for identification and explanation without intent to infringe.

British Library Cataloguing-in-Publication Data
A catalogue record for this book is available from the British Library

Library of Congress Cataloging-in-Publication Data
Names: Hancock, James F., author.
Title: Plantation crops, plunder and power : evolution and exploitation / James F. Hancock.
Description: London ; New York : Routledge, 2017. | Series: Earthscan food and agriculture series | Includes bibliographical references and index.
Identifiers: LCCN 2016042049 | ISBN 9781138285750 (hbk) | ISBN 9781138285767 (pbk) | ISBN 9781315268781 (ebk)
Subjects: LCSH: Plantations—History. | Tropical crops—History. | Agriculture—Tropics—Political aspects—History. | Agriculture and politics—History. | Agriculture—Social aspects—History.
Classification: LCC HD1471.A3 H36 2017 | DDC 306.3/49—dc23
LC record available at https://lccn.loc.gov/2016042049

ISBN: 978-1-138-28575-0 (hbk)
ISBN: 978-1-138-28576-7 (pbk)
ISBN: 978-1-315-26878-1 (ebk)

Typeset in Sabon
by Apex CoVantage, LLC

To my wife Ann for her unwavering support and love.
My life has been greatly enriched by her.

Contents

Figures and Table

Figures

Table

Preface

For decades, I have taught a graduate level course on crop evolution, and have long felt that a book focusing on the crops grown on plantations would be an interesting journey through world history. There are a number of popular and academic books that discuss how the domestication of food crops shaped human civilization, but no one has described how the plantation crops came to dominate world politics and society. These crops were grown for thousands of years by small farmers until world demand for their products grew to the point that European explorers and traders established vast plantations of them for profit. All strongly influenced government affairs and fostered slavery in one form or another.

My initial thinking was that each crop's story would be unique and would stand alone. When I had completed the first draft of the book and began to seek a publisher, I learned that telling good stories about individual crops was not going to be enough. I would have to find a common theme to pull the project together. As one editor put it, the book needed "an over-arching thesis" and "to tell us something beyond the details of each account." Being forced to think about the plantation crops as a whole, I realized that they all did have a common path of development that revolved around innovation, exploitation and war. This was my "over-arching thesis," which I then described in a concluding chapter.

In the next round of submissions, I found that the book had to pass one additional hurdle for publication. I had to find a publisher willing to consider inter-disciplinary studies. My book did not fall cleanly into any specific scientific discipline and this approach confused most editors. Finally, I got lucky and found Tim Hardwick, Senior Commissioning Editor at Earthscan from Routledge. He had worked with me on two previous books at CABI and clearly understood what I was trying to accomplish. Tim proposed a title change that better reflected what the book was about (with alliteration!) and offered a number of valuable suggestions on the book's construction and content. He pushed me to demonstrate in my final chapter how the plantation crops had developed along similar themes that are still playing out today. With Tim as my champion, the book was finally accepted and published.

Introduction

Role of plantation crops in world history

This book weaves together the stories of seven diverse crops that came to be grown in plantations – sugarcane, banana, cotton, tea, tobacco, coffee and rubber. It marries the evolutionary history of each crop with the social and political upheavals associated with their cultivation and trade. From the early 1500s until today, at least one of the plantation crops has always been at the center of international politics.

By plantation crop, I am referring to those crops grown in large farms devoted solely to their culture. They are grown for profit and their products are shipped long distances to market. Of course, there are many more plantation crops than just sugarcane, banana, cotton, tea, tobacco, coffee and rubber, but these are the most widely planted and have particularly long, rich histories. The chapters of the book are ordered by the date of each crop's first domestication.

While each of the plantation crops has its own unique aspects, all developed along a similar blueprint. First, they were grown by small landowners who used their products locally, but eventually worldwide demand grew to the point that plantations were established in distant lands to meet that need. Vast tracts of native vegetation were clear-cut to farm the plantation crops, and slaves or indentured servants were imported to harvest them. These people were brutally treated to keep profits high. The economies of whole countries became tied to the profits of these plantations, resulting in internal power struggles to control the burgeoning wealth. Open warfare routinely broke out between the more powerful countries for dominance in their trade.

The plantation crops have dominated world affairs in myriad ways. Sugar fueled the settlement of the Caribbean and played a pivotal role in the American Revolution. Cotton was at the center of the Industrial Revolution; its demand encouraged the rebirth of slavery in the United States and led to the Civil War. Banana became the province of a few giant personalities, unafraid to overthrow countries with US government help. Coffee started as a drink that sustained religious ceremonies, and then became the center piece of intellectual life across the world. Rubber was first harvested by a few ruthless men in Brazil, only to be smuggled out of the country in order to establish vast plantations in southeast Asia. Tea was the nucleating force that led to the invasion and breakup of the Celestial Empire, China. Tobacco started small as a Native American crop but grew to vast importance as millions of people worldwide became addicted to it.

There is no archeological record of when and where humans first began growing sugarcane as a crop, but it was most likely domesticated about 10,000 years ago in what is now New Guinea. The early farmers surely brought dramatic changes in the sugar content of sugarcane by starting new gardens year after year with only the sweetest individuals. The geographic center of sugarcane cultivation shifted gradually across the world over a span of 3,000 years: starting from India to Persia, it moved along the Mediterranean to the islands near the coast of Africa and ultimately the New World, first in Brazil and then the Caribbean. Sugar barons became rich by brutally enslaving people to manufacture sugar. European traders amassed fortunes by mercilessly delivering slaves in the cramped, suffocating holds of their ships and returning to their home ports with "white gold." After the Haitian slave revolt, many sugar planters fled to Cuba and Louisiana where slavery was still in practice. In the mid-1800s sugar culture spread to British and Dutch Guiana (Guyana), East Africa, Mauritius, Natal (South Africa) and Queensland (Australia). The sugar plantations in all these places depended on indentured workers, and the primary labor source was desperate "coolies" from India.

The earliest evidence of banana cultivation, dating back at least 7,000 years, comes from the highlands of New Guinea. It is hard to imagine what humans saw in wild bananas, as they have only a small amount of fleshy pulp that is filled with flinty, teeth-breaking seeds. The transition from wild to really edible bananas must have involved the gradual selection for both seed suppression and parthenocarpy (the production of fruit without seeds). Banana began in antiquity as a staple food crop, but was transformed in the late 1800s into an important plantation crop. A few industry leaders ("Banana men") established these plantations; they wielded great power by developing all-encompassing, vertically integrated companies and building extensive shipping empires. Vast swaths of Central and South America were deforested and replaced by thousands upon thousands of acres of a single variety of bananas. Countries in Central America became servants of the great banana companies, offering them sweet tax and land deals in return for bribes and infrastructure building. US-supported coups became commonplace whenever host countries "misbehaved" and began to cut into company profits. Workers were exploited and worked nearly to death clearing forests for planting and harvesting fruit. Periodic riots ensued in response to the harsh working conditions. Disease persistently stalked the monocultured fields and to keep it in check, the companies applied pesticides liberally by land and air, sickening both workers and the environment. The banana industry became a showpiece for what unchecked corporate power could unleash.

Cotton had a most improbable origin involving species in two continents that somehow managed to hybridize. The cotton species now grown were domesticated about 6,000 years ago. It is remarkable that people began cultivating cotton, as the wild species produce only sparse amounts of coarse seed hairs that are short and very difficult to spin into thread. The ancient farmers must have imagined cotton's potential, and gradually improved it by painstakingly selecting seed from the most promising plants generation after generation. For thousands of years, India was the center of textile production, supporting a vast cottage industry. This collapsed when Europeans finally learned to make quality yarn in the mid-1700s and print it with lasting, vibrant colors. The subsequent boom in British textile production led to the Industrial Revolution and supported a global empire. Unfortunately, work conditions in the mills became abominable, and sweatshops emerged where hapless immigrants toiled under repressive conditions to sew clothing. Slavery in the US South got new life, producing raw cotton for the British in the mid-1800s, and a textile industry emerged in the northeastern United States. Slavery was eliminated through war and sharecroppers took over the bulk of cotton production. The US textile industry moved south in the late 1800s and riots over pay and working conditions became commonplace.

Tea was first planted in China about 2,500 years ago, after being harvested from the wild for thousands of years. No one really knows when humans started making tea, but it is likely that early Cro-Magnon were chewing raw leaves for thousands of years before the advent of agriculture. Chinese legend has it that the first cup of tea was made by the second emperor of China, Shen Nung, about 5,000 years ago. Tea became a staple of Japanese life in the 1600s and the British fell head over heels for it in the 1700s. The growing international demand for tea pulled a reluctant China into the worldwide trading arena. To get silver to pay for Chinese tea, the British began cultivating and selling opium from India, and when the Chinese emperor balked, the British started the Opium Wars. Tea plantations were established in India and southeast Asia in the 1800s, which soon eclipsed the Chinese trade, and contributed greatly to the political and social upheaval of that country. To find enough laborers to work the tea plantations in India and southeast Asia, the use of indentured servants became the norm. These displaced people suffered appallingly high death rates due to disease, malnutrition and overwork.

Most archeologists believe that tobacco was first domesticated 5,000 to 7,000 years ago in South America. How people came to domesticate the common tobaccos of today is an intriguing puzzle, as they do not exist in the wild. Tobacco cultivation had spread all across the Americas by the time Columbus stumbled on its use in 1492. Europeans were initially fearful of tobacco, but its medical use soon spread and pipe smoking took England by storm in the sixteenth century. The English demand for tobacco lead to a lucrative farming industry in the Virginia colonies in the 1700s, and the importation of slaves to boost productivity. Pipe smoking was replaced by cigarettes in the late 1800s, when machines were invented to roll the cheap, convenient product. It was not long until vast numbers of people across the globe were addicted to cigarettes. In the 1920 and 1930s, many health practitioners became alarmed at the rising incidence of lung cancer in the United States and Great Britain, and research ultimately bore the association with tobacco out.

The Ethiopians were the first people to drink Arabica coffee, perhaps 3,000 years ago. No one really knows when the first coffee beans were used to make a beverage, but the story most often told involves a shepherd named Kaldi whose flock of goats became frisky after consuming beans from a coffee tree. Ethiopia was the only place coffee was grown for about 1,500 years until the 1700s, when a few coffee plants were spirited out of the country, propagated and dispersed to all corners of the tropics. The Dutch established the first large coffee plantations in Java, followed closely by the English in Ceylon. The Dutch paid local chiefs to force their subjects to harvest their coffee, while the British brought in desperately poor indentured servants from India. Production boomed in southeastern Asia for decades until the fields became decimated by the coffee rust. Brazil came to dominate world production in the mid-1800s, a position it still holds today. Central America and Colombia became major players in the late 1800s and early 1900s, and Vietnam emerged as a major producer in the last few decades. The Latin America coffee industry was built on the backs of the Mayans who were forced to harvest the crops for a pittance at best. The global coffee industry has suffered tremendous annual fluctuations in the crop's worth, as production has often out stripped demand.

Rubber was first used by the Aztecs, Incas and Mayans to make a number of objects including balls for their court games. Thousands of ball courts were built across their world between 1500 BC and AD 1521. Columbus was the first European to report on rubber after his second trip to Haiti, but Europeans had little use for it until innovators in the mid-1800s figured out how to use it to make waterproofing, gaskets and tires. To fuel the resulting craze for rubber, native peoples in Brazil and the Congo were forced by ruthless landowners to harvest the sap of natural stands of rubber trees and other latex producing plants. Seeds of rubber trees were eventually smuggled out of Brazil and propagated to establish vast rubber plantations across the tropics of southeast Asia. Indentured workers were brought in to work on these plantations under brutal, repressive conditions. When these plantations were laid fallow by the Japanese during World War II, German and US innovators stepped in and figured out how to make synthetic rubber. To build the necessary factories to produce this rubber, the Germans used Jewish slaves.

Numerous books have been written about the impact of crops on human civilization, but most of them have focused on the crops grown for food. Herein the plantation crops receive their due as major drivers of world affairs. They spotlight the best and worst of human nature, including innovation, exploitation and war.

1 Sugar

Beginnings

There is no archeological record of when and where humans first began growing sugarcane as a crop, but it most likely happened about 10,000 years ago in what is now New Guinea in the South Pacific (Warner 1974; Daniels and Daniels 1993; Hancock 2013). The species domesticated was *Saccharum robustum*, found in dense stands along rivers. The people in New Guinea were among the most inventive agriculturalists in the world and domesticated a broad range of local plant species including not only sugarcane, but also taro, bananas, yam and breadfruit.

The New Guineans probably enjoyed sucking and chewing raw sugarcane long before anyone thought about extracting its sugar. People likely desired a more convenient source of this treat and begin transplanting the sweetest individuals into garden plots close to their homes. Sugar cane was probably also utilized as livestock feed and to build fences and walls. In fact, the New Guineans may have gotten the idea to plant sugarcane when they observed that the canes in their fences and walls often rooted and began to grow.

The early sugarcane farmers surely brought dramatic changes in the average sugar content of *Saccharum robustum*, by starting new gardens year after year with only the sweetest individuals. The sugarcane we grow today is called *S. officinarum*, and it was transformed from the wild *S. robustum* through this human selection. These two species share the same chromosome number, are morphologically quite similar and are sexually compatible. The only major difference between them is the sugar content of their canes.

Early spread

The cultivation of sugarcane moved steadily eastward across the Pacific, spreading to the adjacent Solomon Islands, the New Hebrides, New Caledonia and ultimately to Polynesia (Macinnis 2002). It was the Lapita people, known for their distinctive pottery, who carried sugarcane across the vast Pacific region, reaching Hawaii about 1,000 years ago. They carried with them all the foods and animals domesticated in southeast Asia including taro, bananas, yams, breadfruit, pigs, dogs and chickens.

Cultivation of sugarcane also moved westward into continental Asia, Indonesia, the Philippines, and then Northern India. During this advancement, *Saccharum officinarum* ("noble canes") hybridized with a local wild species called *S. spontaneum* to produce a hybrid, *S. sinense* ("thin" canes). These hybrids were less sweet and not as robust as pure *S. officinarum*, but were hardier and could be grown much more successfully in subtropical mainlands.

It was in India about 3,000 years ago that people first began squeezing the canes and producing sugar (Gopal 1964). The Buddhist scholar, Buddhaghosa, described some of the steps in sugar making. Refining sugar, as we will describe more fully later, is a very laborious process and it is not surprising that it took humans so long to begin its manufacture. The canes were first harvested and cut into small pieces, which were squeezed between rotating and stationary stones. The residue was then boiled to produce crystals, drained and allowed to form sugar cakes.

For a long time, the Indian people kept the whole process of sugar making a closely guarded secret, resulting in rich profits through trade across the subcontinent. It wasn't until the Persian Emperor Darius invaded India in 510 BC that the outside world learned about the process of sugar making. The victors took the technology back to Persia and began producing their own sugar. By the eleventh century, sugar constituted a significant portion of the trade between the East and Europe. Sugar manufacturing continued in Persia for nearly a thousand years, under a revolving set of rulers, until the Mongol invasions of the thirteenth century destroyed the industry.

The Muslim legacy: Mediterranean sugar industry

When the Prophet Muhammad began his holy war to convert the world to Islam in AD 632, his followers simultaneously started an agricultural revolution (Watson 1974; Decker 2009). It all started in their early Persian invasions, when they discovered not only sugarcane, but a long list of crops largely unknown to the rest of the world including artichokes, bananas, coconut palms, cotton, eggplants, lemons, limes, mangos, rice, spinach, sorghum, sour oranges, sugarcane, watermelons, wheat, and yam. In what has been called the Arab Agricultural Revolution, as Muhammad's armies conquered new regions, they introduced this assemblage of crops and in so doing dramatically altered the agriculture of the whole Mediterranean region.

The Arabs also dramatically altered the methods used to grow crops. As S.W. Mintz (1985, p. 25) put it "The Mediterranean's Arab conquerors were synthesizers, innovators transporting the diverse cultural riches of the lands they subjugated back and forth across portions of three continents, combining, intermixing, and inventing, creating new adaptations." Traditionally, crops had been sown only in the late summer to capture the autumn rains. The Muslims began rotating crops and planting throughout the year, allowing for more intense use of the available land. Longer season crops such as winter wheat might be followed by summer sorghum, or crops with shorter growing seasons like spinach, yam and eggplant might be grown back to back in the same year. This multiple cropping depleted the soil of its nutrients more rapidly than single crops, so the Arabs began using manures and other supplements to maintain the soil's fertility.

Year-round planting also required the development of dependable irrigation systems to provide the crops with sufficient water, particularly during dry seasons. To accomplish this, the Muslims restored the old Roman irrigation systems that had fallen into decay and greatly expanded them. They also added new innovations such as the Persian bucket wheel, the water screw and underground water tunnels.

Sugar cane, called the Persian reed by the Muslims, was introduced by them into Egypt in AD 710. Here, sugar refining was taken to a new level, making it whiter and purer by repeated washings and extractions. Egypt became the most sought out source of sugar in the world, reaching its peak production from about 1000 to 1350.

From Egypt, sugarcane was spread westward by the Arabs across Northern Africa, reaching Morocco and Spain in the middle of the seventh century (Galloway 1977). Significant sugar production began in Morocco in the late 800s, reaching its peak from 1000 to 1200. Production of sugar on the Iberian Peninsula began about 900, with its boom starting in 1300 and lasting until 1500. These areas remained under Muslim control throughout this period.

The Arabs also introduced sugar to Cyprus, Crete and Sicily by the early 800s, but significant sugar production did not begin on these islands until well after the Arabs had lost control of most of them. Peak production on Cyprus was from 1300 to 1500, when the island was dominated by the Genoese merchants. The boom years in Crete were from 1400 to 1500 under Venetian rule. In Sicily, the golden years of production were from 800 to 1050, when the Island was still under Muslim control and then again from 1380 to 1520, when it was essentially under Spanish rule.

Sugar use in Europe

When sugar first arrived in the continent from the Near East, honey had been the primary sweetener for thousands of years (Mintz 1985). The oldest evidence of honey use is found in the Arãna caves of Spain, where 12,000-year-old paintings depict what appears to be a person robbing a bee hive on a cliff. The making of mead (honey wine) dates back at least 4,000 years and it was most likely the first fermented drink. Beekeeping probably began about 3,000 years ago.

Europeans did not become aware of sugarcane until Alexander the Great's army stumbled upon it during their conquest of India in the 330s BC. It was first observed by Alexander's admiral Nearchus when he sailed from the Persian Gulf along the Indus River. It is unclear how much he learned about the way sugar was extracted, but it is clear that he had tasted the sweet canes and was aware of its alcohol by-product. Nearchus described sugarcane as "a reed in India brings forth honey without the help of bees, from which an intoxicating drink is made though the plant bears no fruit."

Knowledge about sugar continued to grow in Europe during the Crusades of the eleventh to thirteenth centuries (Aronson and Budhos 2010). As the Christians won new territory, they took over sugar production and began to export it themselves to Europe. Sugar was used primarily as a medicine and to mask the unpleasant taste of other pharmaceutical ingredients. It was recommended by the Greeks and Italians for breaking fever, stanching coughs, soothing chapped lips and curing stomach disorders. The influential theologian Thomas Aquinas deemed sugar a medicine rather than food, making its consumption allowable during the fast.

Sugar became a possession of only the very rich, because it commanded prices as high as that of the other spices from the Far East, such as pepper, saffron and cinnamon. It really wasn't used much in the preparation of foods, but when it was, it was employed in what we would consider overbearing amounts. Meats were literally drowned in heavy syrups with almonds and fruits.

Sugar and the other spices first became widely available to Europeans in the "Champagne Fairs" hosted in a number of different French cities (Aronson and Budhos 2010). People came from all over Europe to attend these fairs, and traders from Italy sold items purchased from Muslims such as sugar, oranges, apricots and figs, exotic dyes, cotton and silk fabrics. The traders also sailed to Syria, where they could obtain spices from India such as black pepper and sugar. These had come to Syria via overland trade routes. These Champagne Fairs remained important until the 1300s.

In the thirteenth and fourteenth centuries, Venice established itself as the main importer and exporter of sugar. In the beginning of this period, raw cane sugar was imported from India and refined in Venice before export to the rest of Europe. Eventually, Cyprus, Crete and Sicily became the major European suppliers of sugar.

Expansion of sugar manufacturing by the Portuguese and Spanish

The Portuguese ultimately took control of worldwide sugar production in the fifteenth century as an economic by-product of their exploration and colonization of the Atlantic Islands along the African coast (Macinnis 2002; Dewald 2004). It all began in the early 1400s, when King John I of Portugal became interested in opening a sea route to the North African Sahara Trade

Routes. He was interested in obtaining gold, as well as slaves, pepper and ivory.

King John I began his African adventures in 1415 by conquering the Muslim city of Ceuta in northern Africa. This opened the door to the caravan trade routes and the spices of the Far East. To get gold, the Portuguese trafficked in African slaves. Most of these were sold in Africa, although 800 to 900 Africans were sent to Portugal each year, supplementing the Muslim ones already held in servitude as spoils of war. Most of the slaves held in Portugal were used as domestic servants; large-scale enslavement awaited the establishment of the Portuguese sugar plantations.

The Madeira Islands were discovered by the Portuguese explorers João Gonçalves Zarco and Tristão Vaz Teixeira in 1419 and the Portuguese rapidly colonized these uninhabited, tree-covered islands. The mountainous terrain required extensive terracing to grow crops. The first exports were cedar and yew wood as the land was cleared, and the dyes dragon's blood, orchil, and woad. Wheat was also briefly a major crop.

In 1420, Prince Henry decided that sugar held the greatest future for his Atlantic acquisitions, and he sent to Sicily for cane plants and experienced sugar technicians. The first sugar was refined in Madeira in 1432 and by 1460 the island was the world's largest sugar producer. Sugar was the main product of the island until the mid-sixteenth century, when it was gradually replaced by wine.

In the fifteenth century, the Portuguese also discovered and developed the unpopulated Azores, Cape Verde Islands and São Tomé, while the Spaniards conquered the already inhabited Canary Islands. Sugar never became important in the Azores and Cape Verde Islands, but the Canaries and São Tomé became major sugar producers by the late 1400s.

Dirty necessity: slavery and the plantation system

Once the Muslims had built a commercial market for their sugar, they began a whole new system of agriculture that relied on masses of people, speed and organization (Watson 1974; Galloway 1977; Decker 2009). The production of sugar was run much more like a factory than a farm. In what is now called the plantation system, the harvested product was almost immediately transferred to the processing plant where a veritable army of workers toiled around the clock until the harvest was complete. Sugar was produced within twenty-four to forty-eight hours of the raw canes' harvest on the world's first assembly lines.

As sugar production became big business, it required a huge capital investment. Large numbers of people were needed to plant, harvest and process the crop. As Peter Macinnis (2002) outlines in his book, *Bittersweet*:

> Fuel was needed to heat the juice, land to grow the cane, and reserves to keep the growers going for 15 months until the first crop matured. They

> needed transportation to get the cane to where it would be crushed, iron hoops and timber staves for barrels, coopers and carpenters, a blacksmith to make or repair tools, animals and animal drivers, people to gather feed for the animals and to grow food for the workers, and much more.
>
> (pp. 20–21)

In addition, a lot of cane had to be cut to get economic yields of sugar. The cane contains about 10% fiber, 75% water and 15% sugar; therefore, to get a ton of sugar, over seven tons of cane must be cut.

In the early years of Muslim sugar production, small peasant-owned farms often existed next to large estates. The estates were partitioned out to "sharecroppers," tenants who paid the owners with a portion of their harvest. Initially in the Crusader states, as well as Crete and Cyprus, land was controlled by feudal lords who drew on the labor of a serf class. The sharecroppers and serfs were not slaves, but they were certainly not free.

It did not take too much more than a century for sugar production to be tied to the business of slaves. It became impossible to find enough labor, and it was much too costly to make large quantities of sugar without enslaving people. To make cheap sugar and maximize profits, the plantation system was developed, where a large slave population was forced to live and work the land and mills. This system would be adapted to other commercial crops such as cotton, indigo and tobacco, but sugar was the first. Sugar and the slave trade would reach its peak when the Europeans began to exploit their newfound territory in the Atlantic, South America and the Caribbean.

Slaves first became "a necessity" in the fourteenth and fifteenth centuries in Crete and Cyprus, when sugar had become big business and the ravages of warfare and Black Death had decimated local populations (Galloway 1977). The slaves came from where ever wars were being fought and included Bulgarians, Greeks, Tartars and Turks.

Figure 1.1 A fanciful Theodor de Bry engraving from *De Americae* (1595), the earliest depiction of the sugar industry in the West Indies

When they began their sugar estates in the Canaries, the Spanish started out by brutally suppressing the native Guanches and forcing them to work in the cane fields (Sokolow 2003); Dewald 2004). When there were too few Guanches left alive to work their fields, they imported African slaves. The Portuguese first exploited imported Guanches and North African Muslims in their Atlantic island plantations, but soon replaced them with the Africans that they were already trading. Thousands of slaves were sold to plantations in the Atlantic islands. By the early sixteenth century, there were a total of 20,000 people on Madeira and 10% were African slaves.

The source of the African slaves used in the burgeoning Portuguese sugar colonies gradually shifted south from the Sahara trade routes to what is now called Angola. In their march down the African coast, the Portuguese had come in contact with kingdom of Kongo in the 1480s, south of the Congo River. The Kongo were ferocious fighters that captured and enslaved the people of the defeated kingdoms. Sadly, the Portuguese became their allies and traded guns for slaves.

The tedium and toil of sugar production

The production of sugar by the slaves was a tedious, dangerous and labor-intensive process, and was often conducted in places with almost unbearable hot, humid conditions (Macinnis 2002; Aronson and Budhos 2010). First came the land clearing, often beginning with the removal of a forest. Then, small pieces of cane, called setts, had to be planted by stooped-over labor. Men and women worked in long rows to prepare the ground and plant the setts.

Once established, the planting had to be periodically weeded by a short-handled hoe, again in a backbreaking fashion, until the cane became rooted and had grown enough to successfully compete with the weeds that shade them out. This work was done primarily by women and boys, who typically worked ten to twelve hours a day under the hot sun, with rats running around their ankles. Rats and cane fields seemed to be common companions. As the cane grew, gangs of workers were sent out to cut and stockpile wood to feed the boiling vats when the harvest began.

The harvest began about fifteen months after planting, when the sugarcane had grown into dense thickets of towering plants, up to thirteen feet tall. Sweating, miserable people working under the torrid sun would cut the cane for hours on end with machetes, taking care to not to slash themselves or others. The cane itself had bladelike teeth along its surface, tearing at their clothes or any skin touching it. Men and women worked in teams, with the men slashing the cane and stripping off the leaves, while the women bundled it in manageable groups of twelve. The teams were often required to generate over 4,000 bundles a day, depending on the speed of the mill and the number of people in the field. The harvest went on continuously for up to ten months depending on local conditions.

The bundles of cut cane were immediately hauled to the nearby mill, often in carts driven by children. Haste was imperative, as soon after the cane was cut, it began to turn woody and its enzymes began to break down the desired product, sucrose. At the mill, the cane would be cut into five- to six-inch pieces and then crushed to free its sap. Crushing was done until the 1300s with a grindstone, but was ultimately replaced with a roller mill by the Sicilians, which consisting of two long, rotating discs that greatly increased productivity. The grindstones were turned by humans or animals, while the rollers were rotated by water or wind.

The crushing went on continuously throughout the day, exhausting the women who most commonly fed the canes into the rolling grinder. An axe or sword was kept close by to remove the limb of anyone who fell asleep while pushing the cane under the rollers.

The crushed cane would then be squeezed in a press to extract as much juice as possible. The juice was then poured into vats, and boiled down in repeated steps. Sweating slaves watched continuously over the process, periodically removing foams of debris and by-products with wooden ladles. The huge hot fires raged, demanding a constant input of wood. Woe to the exhausted person getting too close to the fires or the boiling cauldrons. When the steaming mass was deemed thick and rich enough in sugar, it was poured off and allowed to cool to form dark-brown, sugar cakes. To produce snow white sugar, the process had to be repeated over and over again.

Once the harvest began, the pace at the factory must have been frantic. Masses of people had to be coordinated to do their individual jobs like clockwork. A delay at any stage of the process could reduce the yield of sugar; therefore, time literally was money. Tight, rigid discipline was maintained, with people laboring around the clock until all the cane was harvested. The mills worked brutal shifts of about eighteen hours, from 4:00 in the afternoon to 10:00 the next morning. It was likely that the workers were pushed to the absolute limits of their endurance to keep the sugar factory operating smoothly.

American expansions – Brazil and the Caribbean

In the sixteenth century, the center of sugar production began to shift to the Spanish-controlled Caribbean, first in Santo Domingo, and then to a smaller extent in Cuba and Puerto Rico (Macinnis 2002; Aronson and Budhos 2010; Parker 2011). Christopher Columbus had introduced sugarcane to the region on his second voyage of 1493. The Spanish were still much more interested in finding gold and silver, but they found the profit of sugar too enticing to pass up. They imported skilled sugar masters from the Canaries in 1515 and sent their first shipment of sugar to Europe soon after.

The Portuguese discovered Brazil in 1500 and it did not take them long to begin planting sugarcane there (Burns 1993). They first focused on trading with the native Tupi for Brazilwood, which was used to make a red dye, but they began to establish colonies in earnest when the French starting taking too much interest in their new territory. The first sugar plantation was established in 1518 and by the late 1500s Brazil had become the leading supplier of sugar to the European markets. Brazilian sugar production reached its peak in the 1620s in the Pernambuco and Bahia regions, at about 15,000–20,000 tons a year.

A large percentage of the early settlers in Brazil were what were called "New Christians," descendants of former Jews that were persecuted during the Inquisition. The burgeoning sugar industry of the Portuguese was first largely financed by Dutch merchants, but the Dutch eventually established their own colony in northeastern Brazil. There they were active sugar producers from 1630 to 1654, until they were expelled by the Portuguese.

When the Dutch were forced to leave Brazil, they took their money, expertise and shipping to the Caribbean, which was closer to Europe and being actively colonized by the British and French. Within thirty years, the Caribbean took over dominance of worldwide sugar production, driving the price of Brazilian sugar down by two-thirds and severely reducing their exports.

The Brazilian sugar industry found its stiff competition, first from the tiny island of Barbados, and ultimately from a hodgepodge of islands controlled by the British, French and Dutch (Macinnis 2002; Parker 2011). As sugar production spread across the Caribbean, it fueled a massive growth in slavery, stimulated almost continual warfare and piracy and made many people massively rich. The size of the slave population grew to such an extent that planters, fearing slave uprisings, resorted to ever-increasing brutality to control their ranks.

Originally settled in 1627, Barbados had become one massive sugar factory by the 1640s, dominated by a handful of large landowners. They learned the technology from the Dutch whose sugar production on the Guiana coast had been thwarted by the Portuguese. The life there was hard and fast, with lives being shortened by rampant disease and booze, but the hardiest (and luckiest) amassed vast sums of wealth and lived opulent lifestyles patterned after that of British royalty.

By the late 1600s, sugar production became important on a number of Caribbean islands including the British-controlled Antigua and Nevis, French-controlled Martinique, Guadeloupe and St. Dominique (now Haiti), and sections of St. Kitts controlled by the French and British. Dutch merchants had amassed great wealth ferreting goods to and from the Islands. Ultimately, British Jamaica became the crown jewel of Caribbean sugar production after a long and difficult settlement period.

The boom of sugar use begins

By the middle of the eighteenth century sugar had become an integral part of British life, tied closely with the introduction of tea, coffee and chocolate (Mintz 1985; Macinnis 2002; Aronson and Budhos 2010). The consumption of tea with meals (and on break) was now widespread, and much more sugar was being consumed by the working class and poor than the rich. Cookbooks were being printed that described how to make sugar sculptures and sugared custards, pastries and puddings. Common folk used sugar or molasses (treacle) to flavor breads, biscuits, tarts, buns and porridges, including what was called "hasty pudding." By 1850, about one-fifth of the calories eaten by the British were in the form of sugar.

A new product called rumbullion (or later rum) was also developed about this time from sugar processing by-products and provided the world with a new source of cheap liquor. Rum production proved to be a boom for sugar producers, as it provided a stable, easily stored source of income and was at the ready to drink for merriment or solace. In 1698, only 207 gallons of rum was imported to England, but by 1771 it was over two million gallons!

Rum became a particularly important part of the sailor's life. Before its invention, wine, beer and brandy were given to sailors to warm their bellies and bolster them for battle. When Caribbean rum production really ramped up, it became the favored drink of the navies, being cheap and readily available. The sailors of the seventeenth and eighteenth centuries commonly were given a daily allotment of watered-down rum called "grog," one-pint rum to one-quart water.

Conquest and slavery in the Americas

When the Portuguese arrived in Brazil in the early 1500s, they discovered that the area was already inhabited by the Tupi, mostly naked hunter-gatherers and warlike cannibals (Macinnis 2002). The Portuguese quickly set about trying to subjugate them to work in their mines and harvest their sugarcane. They did this in two ways. Jesuit missionaries tried to convert them to Catholicism and recruit them to live in colonial villages to work the farms. Expeditions of adventurers called "bandeirantes" also enslaved thousands of Tupi as they pushed toward the interior searching for precious metals and stones. The Church had ruled that any "savages" that refused to be converted to Christianity could be sold as slaves.

The Tupi proved to be poorly adapted to the routine, sedentary lifestyle of farming and were particularly uncooperative slaves. They were also subject to Western diseases and found it relatively easy to run away and hide in the dense forest. Neither a conversion to Christianity nor brutal treatment worked to make them proper, docile farm laborers. The all too natural solution to this labor problem was to turn to African slavery, a system the Portuguese had already employed in their Atlantic sugar plantations off the coast of Africa. By the mid-sixteenth century, African slavery predominated on the sugar plantations of Brazil, although the enslavement of the indigenous people continued well into the seventeenth century.

Like Brazil, the Caribbean islands were also inhabited when the Europeans arrived (Sokolow 2003). The most numerous occupants were the Arawaks (or Tainos) who were found in most of the Greater Antilles (Cuba, Jamaica, Hispaniola and Puerto Rico), and the Caribs who resided in the Lesser Antilles. Both groups were fishermen and hunter-gatherers and had done little farming. Almost immediately, the Spanish began enslaving the Arawaks, forcing them to work their mines and plantations. The Arawaks initially fought some spirited battles to maintain their freedom, but the Spanish soldiers proved too much for them. Within a few decades of the arrival of Columbus, this whole proud people were almost extinct, dying in droves due to brutal, cruel treatment, loss of heart and susceptibility to the diseases brought by the Europeans.

The fiercer Caribs did comparatively better than the Arawaks. Large groups of them were never enslaved, surviving for centuries on some of the more desolate islands, periodically raiding and taking captives from the French and English islands. Still, the rifles of the Europeans and their diseases slowly but steadily decimated their numbers, until all that was left by the late 1700s was a couple of thousand souls hidden in Dominica's protective mountains.

As in Brazil, the Caribbean sugarcane planters quickly became dependent on enslaved Africans as their workforce. However, they also made use of white indentured servants, who to a large extent were political prisoners. Undesirables of all shapes and sizes were directed to the Caribbean including debtors, small-time criminals, political and religious problem-makers, labor organizers and a very large group of Irish revolutionaries. In theory, they were to be given land at the end of their indentured period and help fill up the New World. They were also desperately wanted as white faces to help balance out the growing number of African slaves. In the end, they proved to be headstrong, difficult to control laborers and were rapidly outnumbered by the imported African slaves. Over 250,000 African slaves were brought into Barbados from 1701 to 1810 and 650,000 to Jamaica (Parker 2011). The number of whites in Barbados never exceeded about 30,000 in the sixteenth and seventeenth century, while in Jamaica they were never more than 20,000.

Human cargo – the Middle Passage

The European use of African slaves had actually begun in the mid-fourteenth century when the Portuguese began transporting slaves to Europe. By 1455, slave labor was common in the Portuguese Algarve and 10% of the people in Lisbon were of African origin. But this was just a small prelude of what was to come in comparison to the level of human trafficking to the Americas. The first major buyers of slaves in the New World were the Spanish on Hispaniola in 1526 after they had essentially extinguished the Arawaks. The next major buyers were the sugar plantations in Brazil beginning about 1560, and then the sugar plantations in the Americas in the second quarter of the 1600s.

The arrival of slave ships in the Americas continued unabated for more than 300 years (Macinnis 2002; Klein 2010; Anonymous 2013). The majority came from western Africa, where the Portuguese had established trading outposts. The horrendous journey that the slaves took from Africa across the Atlantic to the Americas was called the "Middle Passage." The horror

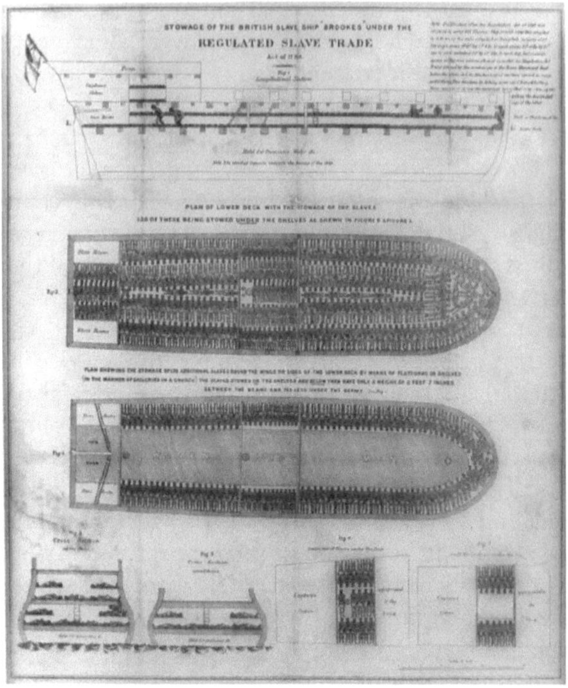

Figure 1.2 Stowage on the British slave ship *Brookes* in the late 1700s

Source: US Library of Congress.

and misery the slaves must have felt is indescribable. These were people who had often been captured far away, marched for long distances and held in holding pens until a slave ship arrived. They were packed naked in the holds of ships, like sardines side by side, chained for long periods of time with little room to move. The air around became stifling, filled with sounds of misery and the stink of sweating bodies, excrement and vomit. And the voyage took forever, the fastest being three weeks, but more commonly dragging on for over two months. An average of 10%–15% of those who were jammed onto the ship didn't survive the journey and were heartlessly tossed into the ocean.

There were essentially two major slave trading routes from Europe to the Americas that followed wind and ocean currents: a northern one to the Caribbean and a southern one to Brazil. These routes evolved into what historians have called triangles of trade, networks whereby manufactured goods (mostly textiles) were shipped from Europe to West Africa, slaves from Africa to the Americas, and ultimately sugar and rum back to Europe. Colonial New England got into slave trade in the 1700s and started its own triangle, shipping rum to Africa, slaves to the Caribbean and sugar and molasses back to North America. New England merchants also traded directly with the Caribbean transferring flour, bread, corn, salted beef and wood for sugar, rum, molasses, limes, cocoa and ginger.

The quantity of slaves that were moved to the New World was simply appalling (Klein 2010; Anonymous 2013). At least 12.5 million unfortunate souls were shipped to a life of hell in the New World. Six European imperial countries were involved in the slave trade from the 1500s to 1800s: Portuguese (5.5 million), English (3.3 million), French (1.4 million), Spanish (1.1 million), Dutch (0.5 million), and Danish (0.1 million). The colonists in New England also got into the act, transferring and selling over 300,000 Africans. Most of the slaves wound up on the sugar plantations, with 70% going to Brazil and the British Caribbean. Only 4% of the slaves wound up in North America. The slave populations in North America ultimately became self-sustaining through reproduction, while those unfortunate souls on the sugar plantations had to be regularly replenished due to extreme mortality rates. The vast majority of the slaves in South America and the West Indies did not survive for more than a couple of years, before disease, brutal treatment and overwork killed them.

Unthinkable brutality and slave uprisings in the Caribbean

As slave populations grew in the West Indies, the white owners became increasingly concerned about slave uprisings. It was clear to them that with the proper spark, the slaves enormous numerical superiority would make them a formidable force. Rather than treat the slaves humanely, in an attempt to placate them and make them less likely to rebel, the owners took the alternate approach and resorted to almost unthinkable brutality. Their goal was to keep the slaves too exhausted and fearful of reprisals to rebel. The British owners also maintained a strong militia that could rapidly respond to rebellions, and they built their homes like fortresses to withstand long periods of attack.

What was done to try and keep the errant slaves docile was simply beyond belief (Aronson and Budhos 2010; Parker 2011). It was common to have other slaves whip them repeatedly with dozens or even hundreds of lashes, and then rub hot pepper into the wounds to maximize agony. Sugar or molasses was sometimes placed on the lesions to encourage insect agitation. Slaves caught chewing cane were fed the excrement of other slaves and forced to swallow it. The most recalcitrant were chained to posts and left to starve, their ears and noses sometimes cut off and fed to them. The very worst offenders were burned alive, often starting at the feet and moving upwards slowly. Female slaves were subjected to repeated rape by their overseers, satisfying themselves whenever the spirit moved them.

And the slaves did repeatedly revolt, with mixed success (Parker 2011). Probably the most complex and sophisticated plan of revolution was concocted in Barbados in 1692. Enslaved leaders from twenty-one plantations put together a plan where they would overrun their individual plantations and form regiments of foot soldiers and horsemen that would march to Bridgeman and take over the arsenal. Irish co-conspirators would get the "garrison drunk" to facilitate its capture. It would begin when the island's white militia departed to fight the French in Guadalupe. The plan fell apart a few days before it began when a couple of slaves were overheard talking about the plan. They were tortured until they started naming names and ultimately hundreds of slaves were arrested. Many of these were exported to other islands, but a high percentage were hanged, burned alive, castrated, and kept in chains until they starved. What made this plan particularly disconcerting to the white plantation owners was that it was developed by the most trusted group of slaves, the "creoles," which had been born on the island and were the most skilled laborers including carpenters, overseers, blacksmiths and boilers.

In the late 1600s and throughout the 1700s, the rough interior terrain of Jamaica provided a rich base of operation for escaped slaves (called maroons) to raid plantations and terrorize the white owners. One community called "Nanny Town" had thousands of occupants and conducted so many raids that travel into the northern parts of the island became extremely hazardous. Whole parishes were abandoned by whites. In what were called the Maroon Wars, British regiments eventually destroyed Nanny Town in 1734, but most of the occupants simply dispersed into the wilderness and continued guerrilla activities.

On the western side of Jamaica, there were also a large number of escaped slaves, many living in a place called Cudjoe's Town. Their leader was initially pragmatic and tried to minimize contact with whites, but the British could not stand the idea of free slaves and repeatedly attacked their settlements. Thus began a long period of raids, retaliations, ambushes and reprisals. Finally, a treaty was reached in 1739 where the maroons were given 1,500 acres to grow food (not sugar!), if they would build and maintain roads connecting the white areas, return future runaway slaves and aid in the event of foreign invasion. Eventually, a similar deal was cut with other groups across the island.

The most significant slave uprising occurred in Jamaica in 1760, in what was called Tacky's Revolt. A group of about 100 slaves from the north central coast overthrew their overseers and raided an arsenal for weapons. They then headed south, killing any whites they encountered, burning buildings, destroying crops and picking up recruits. Their numbers grew to about 1,000 and stimulated uprisings all over Jamaica. Ultimately, it was a maroon who hunted down and killed Tacky, and Cudjoe's followers mopped up the remaining resistance. Overall hundreds of blacks and sixty whites were killed. In the aftermath, at least 500 of the participants were exported to Honduras, and another 100 were brutally tortured and executed in all the typical, vile ways.

Slavery in Brazil

In most regards, slaves on the sugar plantations in Brazil were no better off than those in the Caribbean, being brutally suppressed and often tortured (Williams 1930; Schwartz 1985). Flogging with a cowhide whip was the most popular punishment for minor transgressions to minimize time away from the field. Family units were commonly broken up, as members were cruelly sold and shipped away. The large plantations all had prisons, with shackles, stocks and thumbscrews for extracting confessions.

The workday was essentially sunrise to sunset, as long as seventeen hours during high summer. However, most Brazilian slaves were given a few days off every month or so, and they had Sundays and other religious holidays off. Many had their own garden plots. Slaves could also buy their freedom, albeit against stiff odds. Most slaves were baptized into the Catholic Church, often before they were loaded onto the slave ships. Brazilian law required that they be baptized within a year of their arrival and in most of the large plantations there were chapels where they attended Mass.

Regardless of these minor freedoms, suicide was not uncommon among the slaves and some even killed their children to release them from future bondage. Many simply ran away, as the wilderness was often close at hand. In her paper on slavery in Brazil, Mary Wilhelmine Williams describes how escaped slaves were vigorously hunted down with bloodhounds by black *capitão do mato* (captain of the woods), who were paid by the distance traveled. Recaptured first offenders were flogged, while the repeat offenders were made to wear an iron collar around their neck with two prongs projecting upward. If they had stolen property, one of the prongs had attached five fingers of iron. The most recalcitrant slaves were chained to logs or other heavy weights, which they had to drag along with them as they worked the fields.

Many slaves still escaped from the coastal sugar plantations into the mountains outside of Portuguese control (Aronson and Budhos 2010). These escapees, also called "maroons," formed cities with churches, blacksmith shops, potteries and meeting houses. In the largest maroon kingdom called Palmares, 20,000 to 30,000 people lived a lifestyle that mixed cultures, combining European dress with African accents, building European-style churches but worshipping African gods, and eating multicultural stews of African yams, European spices and local maize. Their greatest leader was Zumbi, a brilliant military strategist, who vigorously defended his homeland against Portuguese raids. He is now considered one of the heroes of early Brazil.

Sugar wars

The most productive years of sugar production in the Caribbean coincided with a tumultuous period of European politics, when France, England,

Spain and the Netherlands were continually at war in various combinations. These wars were fought for supremacy of the seas and dominance in worldwide commerce. The Spanish Empire was in the middle of its long slow decline and the other European powers were scrambling to take over control.

All the European conflicts spilled over into the Caribbean (Parker 2011). The first real trouble started in 1651 when the British passed their First Navigation Act, an attempt to cripple the lucrative Dutch trade with her Caribbean colonies. This act was largely ignored in the docks of the Caribbean islands, but the English Navy and privateers were given free rein to confiscate Dutch cargos in the Caribbean and elsewhere. The British vigorously wrestled with the Dutch for decades until they succeeded in making London the center of world commerce rather than Amsterdam.

From the mid to late 1600s, war continued to storm at regular intervals across the Caribbean. When Oliver Cromwell declared war on Spain in 1655, the English tried to conquer Hispaniola but failed miserably; to save face they invaded and took control of sparsely inhabited Jamaica. English buccaneers camped on the northern shores of Hispaniola were sanctioned to raise havoc among the Spanish in the Caribbean and encouraged to bring their prize booty into Port Royale, Jamaica for sale and reshipment.

After the French declared war on England in 1666, they wrested all of St. Kitts from English control, plundered Antigua and overran Montserrat. All these trophies were ultimately returned by treaty at the end of the war. During the Nine Years' War (1688–1697), France gained the western portion of Spanish Hispaniola (Haiti), St. Kitts again traded hands between England and France, and the English made unsuccessful invasions on Guadalupe and Martinique. The French also invaded Jamaica and managed to introduce the yellow fever virus to the island, but were not able to maintain political control.

As part of Queen Anne's War in the early 1700s, the British took back control of St. Kitts, but subsequently lost it to France along with Nevis and Montserrat; the British also tried to take Guadalupe but failed. During the Seven Years' War, the British finally got control of Guadalupe, along with Martinique, Grenada and Havana, Cuba, while British colonials from North America captured Dominica. The French gained control of St. Vincent and the Grenadines during the American Revolutionary War and regained control of Granada. While the revolutionary wars raged across France, the British captured Haiti and Martinique.

From the Spanish-dominated period to the final supremacy of the British, pirates and privateers ran amuck across the Caribbean, making vast sums of money on confiscated sugar and slaves, and any other cargo they could get their bloody hands on (Parker 2011). One of the most notable was Sir John Hawkins, who drove the Portuguese and Spanish mad during the mid-1500s along all their trade routes. He confiscated and traded all kinds of cargo, and may have been the first Englishman to deal in slaves.

Part of the reason why the Jamaican sugar industry was slow to develop was that buccaneering was considered to be an easier and more exciting way to earn a living. British buccaneers living along the undeveloped coast of Jamaica raged throughout the Caribbean, making their living on requisitioned goods. Not to be outdone, French privateers from Hispaniola hassled Jamaicans, often taking them prisoner, torturing them and robbing them of everything they had of value.

A large group of Anglo-American sailors and privateers, left unemployed after Queen Anne's War, took to piracy in the Caribbean, the eastern US seaboard, African coast and Indian Ocean. In this period called "the Golden Age of Piracy," they were joined by a motley assortment of buccaneers from all across the maritime world.

Sugar politics and North America

The importance of the Caribbean sugar plantations to the European economy had far-ranging effects on North America (Parker 2011). By the 1700s, sugar was the most important internationally traded commodity and was responsible for a third of the whole European economy. The English islands also produced considerable quantities of rum that were consumed by British drinkers all across the world, while the French shipped vast quantities of molasses to the British American colonies, where it was made into rum and sold throughout the continent.

The epicenter of the rum business in North America became the colony of Rhode Island, which had been actively trading with the sugar islands since the mid-1600s, dealing surplus agricultural production (food and animals) for specie, sugar, molasses and rum. Merchants in Newport, Rhode Island, as well as Boston also became active in the African slave trade, often using rum to pay for humans. The burgeoning colonial commerce in slaves, sugar and rum stimulated a major shipbuilding industry in New England.

To grab at least some of the revenue passing between the French and British colonies, Parliament passed the Molasses Act in 1733, imposing a duty on molasses imported from outside the British colonies. The colonial distillers, of course, cried bloody murder, fearing it would dig deeply into their profits; however, the Act was only weakly enforced and had only minimal economic ramifications. Even so, the act was considered "taxation without representation" and was one of the first steps leading to the Revolutionary War.

In the peace negotiations after the Seven Years' War, far less interest was paid among the European combatants to the ownership of the Canadian territories than the Caribbean sugar islands. Many British felt that they should hold onto the conquered French islands for their potential tax revenues, while others feared a glut in British sugar production would depress prices. Still others simply wanted the French out of North America. Ultimately, the English bargained the return of a few of the sugar islands for all of eastern

Canada, in essence a trade of a small piece of their sugar profits for a major portion of the northern fur trade.

In 1764, to replenish their war-ravaged coffers, Parliament passed the Sugar Act to replace the expiring Molasses Act. It placed a duty on sugar produced in the Caribbean islands and the import into the colonies of a wide range of products including indigo, coffee, wine, silks and linen cloth. While the weak Molasses Act had been designed to collect revenue solely from molasses producers outside the British sphere, the Sugar Act encompassed all British-generated sugar and was to be rigorously enforced. Now that the threat of attack by the French and Indians was gone in North America, the colonies were enraged by this new duty and actively resisted it. To a large extent, the Sugar Act was the straw that broke the camel's back and pushed the colonies to declare independence.

When the northern colonies did declare their independence in 1776, the subsequent war was really fought by the British on two fronts, North America and the Caribbean. None of the British sugar islands desired independence, but the Caribbean sugar industry was simply too valuable to be ignored. It was a much more important component of the British economy than the northern colonies.

The British had to maintain a strong presence in the West Indies for several reasons. The Caribbean islands had become a major supplier of covert war supplies to the northern colonies. The French were once again attacking the British-held sugar islands and had their eyes on Jamaica. The Continental Congress had also unleashed a mass of privateers that were seriously impacting on British shipping all over the world.

The French, smarting from the defeats of the Seven Years' War, were more than happy to help the colonists out. Their aid began in a covert operation where they shipped a massive amount of war supplies to Martinique, which were retrieved by the colonists. They also increased their holdings in the Caribbean by capturing the British Leeward Islands of St. Kitts, Montserrat and Nevis. Other supplies to the rebels moved through the Dutch island of Statia, a tiny volcanic rock in the vicinity of Puerto Rico. Thousands of ships from the enemies of Britain brought the Americans war supplies in exchange for tobacco, indigo and other products. The Spanish also smuggled arms to the colonists, first through New Orleans and the Mississippi River and later through their colonies in the West Indies. The Spanish were not so enthusiastic about supporting the rebels as the French, but they still liked the idea of English killing English to distract them from acquiring the remaining Spanish New World possessions.

As the Revolutionary War progressed, illegal trade between the Northern colonies and West Indies became rampant. The simple truth was that the West Indian planters desperately needed American products, as they produced only sugar and not food. Sugar and molasses from Jamaica and the Lesser Antilles were shipped to the northern colonies, while the Americans returned cattle, horses, flour, fruits and vegetables, lumber, salted fish and barrels.

In spite of the illegal trade, food became more and more scarce in the Caribbean as the war progressed, and starvation became a real threat. The British fleets and soldiers defending the islands made matters even worse by taking what they required of available provisions. Slave revolts began to break out all across the islands, exasperated by starvation. The most widespread uprising occurred when it became known that a British regiment was going to be sent from Jamaica to New York in 1776. Slaves burned and pillaged numerous plantations in Jamaica, St. Vincent, Dominica, St. Kitts and Tobago, and almost completely destroyed the city of Basseterre in St. Kitts.

With all the smuggling, harassing privateers, French encroachments and slave uprisings, Britain had no choice but to maintain a strong force in the Caribbean during the Revolutionary War. It is likely that this attention aided the northern colonists greatly in winning their independence. Almost immediately after the French recognized the United States, the British Prime Minister Lord North made securing Britain's Caribbean colonies the country's number one priority. The British commander in North America, Clinton, was ordered to send a total of 13,000 troops to the Caribbean, even at the threat of losing Philadelphia and New York. Rodney was sent to the West Indies with a large navy to stop the smuggling from Statia and to thwart a major effort by the French Navy under De Grasse to conquer Jamaica. Through these efforts, the British were able to keep control of their precious sugar islands, but they likely lost their northern colonies in the bargain.

Freeing the slaves in the British West Indies

Even though slavery had long been outlawed in Britain itself, it was allowed to continue unabated in the Caribbean until the early nineteenth century (Macinnis 2002; Aronson and Budhos 2010; Parker 2011). Far too much profit was being made to allow its abolition and the public was pretty much kept in the dark about the hideous practices used to produce their sugar.

The issue of slavery got its first widespread attention in a notorious court case in 1781 when Luke Collingwood, the master of the slave vessel Zong, asked his insurance company for compensation for 132 slaves he had thrown overboard. He claimed that he had to sacrifice them to save enough water to get the rest of his slaves and crew to port. The claim proved invalid when it was discovered by the court that he had in fact arrived in port with plenty of water and other goods remaining. What he had really done was schemed to gain compensation for the sick and dying, who would not have made it alive to his destination. Sadly, the issue of the brutal murder of the human beings did not come up, only their potential value.

The long battle to eliminate the slave trade was vigorously led in Britain for decades by Thomas Clarkson. While a student at Cambridge University in 1785, Clarkson had written an award-winning essay railing against slavery, and realized he must devote his life to the abolitionist cause. When Clarkson began his campaign, the sugar-mad public knew very little about the enslaved people who produced the sugar. They had no idea that the sweetness that they now craved came from the blood and sweat of an oppressed people. Clarkson and the other abolitionists figured that if the British public knew the real story, they would clamor to have the practice stopped. They hammered the public with tales of the atrocities committed on the slave ships and sugar fields, and called for a public boycott of the bloodstained product.

Besides the work of the abolitionists, many other activities raised the public's awareness of the godlessness of slavery and thus put pressure on Parliament to end the practice. A former slave named Olaudah Equiano published an autobiography in 1789 outlining his life as a slave. He vividly described the brutal realities of the slave trade and the cold-blooded treatment by the masters, and showed drawings of the shackles, handcuffs and other devices used to subjugate and torture. In a massive public relations effort beginning in 1792, the East India Company conducted an antislavery campaign to support their sugar produced in India by freemen. They pounded out the message that if you bought sugar from the West Indies you were killing slaves. The company proposed that a slave produced about 450 pounds of sugar before he died, so a family consuming five pounds of sugar per week was killing a slave every twenty-one months. This was an exaggeration, but hundreds of thousands of British subsequently boycotted sugar from the West Indies and paid a large premium for sugar from India.

The great British reformer William Wilberforce was the first to begin fighting slavery in the British Parliament, encouraged by Clarkson and William Pitt. In 1789 he gave his first speech condemning the "bloody traffic" and introduced his first bill to end the slave trade in 1791. Unfortunately, it was roundly defeated, but he continued to fight on. Sugar barons who had moved from the West Indies to Great Britain proved to be formidable lobbyists and fought hard against abolition efforts in any way they could. Finally, in 1807 Parliament was ready to abolish the slave trade. Contributing greatly to the bill's passage were eyewitness accounts from soldiers who had fought in the West Indies and had seen the torture of slaves firsthand.

The other slaving countries slowly followed Britain's example, often under active pressure of the British. The United States abolished the practice in 1808, the Netherlands in 1814, and Spain and Portugal in 1818. France first abolished the slave trade in 1794, only to have it reinstated by Napoleon in 1802. Final elimination came in 1818. Spain abolished slavery in all its colonies except Cuba, Puerto Rico and Santo Domingo in 1811, and fully stopped trading slaves in 1817 when Britain paid them £400,000. Brazil was the longest holdout, waiting until 1831 to abolish the practice, although the law was little enforced until 1850.

There was still much left to be done, as slavery itself was still legal in the Caribbean. It wasn't until 1833 that slavery itself was abolished by Parliament, and only with many unsatisfactory concessions. Only blacks under the age of six were directly freed, while those older had to serve an unpaid apprenticeship for six years. The slave owners were also to be compensated at the tune of £20 million – 40% of the whole government's budget. This was an unworkable system and slavery was finally fully abolished in the British Empire in 1838.

Slave independence in St. Dominique

By the 1700s, the center of sugar production had moved to St. Dominique, the French half of Hispaniola. Thousands of sugar plantations now dotted its landscape and it had become the richest sugar island. This dominance would literally go up in flames at the end of the century when the slaves successfully rebelled and established a free nation (Aronson and Budhos 2010; Parker 2011). This freedom was won by defeating the armies of Europe's leading powers Great Britain and France.

The rebellion began on August 14, 1791, at Alligator Woods in the northern part of the colony. Here a group of slaves pledged to instigate a bloody revolution to overthrow their white owners and destroy the sugar-making capacity of the Island. Within a matter of weeks, 100,000 slaves were in open rebellion and laid waste to the French colony. Hundreds of cane fields were burned to the ground, their supporting mills destroyed and dozens of warehouses demolished. Any whites that were captured were tortured and slaughtered. The rampage continued for the next two years until the ex-slaves had almost total control of the colony. The French government finally gave up and officially give the ex-slaves their freedom. A man named Toussaint became the de facto ruler of St. Dominique, calling himself "L'Ouverture" (the opening).

But the struggle for freedom was not nearly over, as the British now decided to invade the island and re-enslave the freedom fighters. They were fearful that this revolution would be the spark that ignited the whole West Indies, and they were again at war with France. With help from the Spanish who still controlled Santo Domingo, the British began their onslaught in the fall of 1793, expecting a rapid victory. This proved to be a fatal miscalculation, as the ex-slaves led by Toussaint proved to be excellent, organized fighters. The British also suffered greatly from the standard array of tropical diseases. A broken British force gave up in October 1798 and returned home. Toussaint led an invasion of Santo Domingo in 1801 and freed the slaves there.

By now, Napoleon had taken over power from the bloody revolutionaries and he turned his eyes toward France's struggling sugar empire. He quietly decided that the slaves should no longer be free and sent a huge expeditionary force of soldiers to St. Dominique under his brother-in-law to retake the colony. The French did manage to capture Toussaint in 1802 by reneging on a promise of safe conduct to a negotiating table, but after months of furious fighting they were unable to defeat the gallant local fighters. Almost 50,000 French soldiers died in their futile attempt at re-establishing

Figure 1.3 Depiction of Toussaint accepting the constitution giving him control of Hispaniola

Source: US Library of Congress.

control. On January 1, 1804, St. Dominique was declared independent and became Haiti, the second nation in the Americas (after the United States) to free itself from European control. Fittingly, Haiti was the original indigenous Arawak name for Hispaniola.

The Founding Fathers of the United States wrestled mightily over the concept of Haiti. The United States was founded on the principle of freedom but power was actually restricted to only white, landowning males. There was widespread fear that a similar rebellion could occur in the southern United States among the hundreds of thousands of slaves held in captivity. While John Adams gave the Haitians nominal support during their struggles for freedom, his successor Thomas Jefferson completely refused to acknowledge the newly emerged nation. An embargo was approved by Congress in 1806 on all trade except with French-controlled parts of the Island. The recognition of Haiti stood in limbo in the United States for over sixty years, waiting until Lincoln acknowledged it in 1862 during the Civil War, just before his landmark Emancipation Proclamation.

The sugar industry post-slavery

After the Haitian revolution, many sugar planters fled to Cuba and Louisiana (Macinnis 2002; Aronson and Budhos 2010). Cuba soon became the world center of sugar production, while Louisiana became the sugar bowl of the United States.

The terrible oppressive plantation system of the Caribbean was unfortunately transferred to Louisiana, where slavery still existed. Because it was cooler in Louisiana and the production season was shorter, the enslaved were driven even harder than in the Caribbean. The life expectancy of slaves in the Louisiana sugar fields was so brief that new blood had to be continually imported to harvest the cane. This was in contrast to the rest of the southern United States, where slave populations had actually grown to such an extent that they were in surplus. Political battles began to rage over taking the surplus into the western territories of the United States – one of the critical events leading to the Civil War.

When the Civil War began and cut off sugar supplies from Louisiana to the North, Hawaii became the new sugar production center for the United States (Aronson and Budhos 2010). Sugar cane had actually arrived in Hawaii in prehistoric times and was being grown there by the indigenous people long before Cook's discovery of the islands, but modern plantations were not established until the mid-1800s. One of the first plantations was started by the German immigrant Herman A. Widemann, who was actually sympathetic to the Confederacy.

The sugarcane in Hawaii was harvested and processed by indentured workers. The primary ethnic groups exploited were first the Chinese in the 1850s, followed by the Japanese in the 1880s and the Filipinos at the end of the century. The various ethnic groups were systematically replaced by the plantation owners when they became entrenched and began demanding higher salaries. This repeated shift in the origin of laborers has made Hawaii one of the world's hot spots of ethnic diversity.

Numerous other sugar centers cropped up across the globe in the mid to late 1800s including British and Dutch Guiana (Guyana), East Africa, Mauritius, Natal (South Africa) and Queensland (Australia). The sugar plantations in all these places depended on indentured workers, and the primary labor source was India. Marc Aronson and Marina Budhos in their book, *Sugar Changed the World*, report that over one million Indian indentured workers went to sugar plantations from 1835 to 1917, 450,000 to Mauritius, 150,000 to east Africa and Natal and 450,000 to South America and the Caribbean.

The Indian people recruited, known derogatorily as "coolies," were desperate, starving people who had no other opportunities available to them. They agreed to travel away to work, even though most were traditional Hindus that were forbidden by religion to cross the kala pani (black water). By making the journey they would became defiled, and would not be welcomed back. Local recruiters would sign up large numbers of people, herd them onto ships and then hold them together in cheap, coastal lodgings until bids were received for their employment. Later, this system of recruitment evolved into the "kangani system," where plantation foremen would be sent back to their home villages to recruit new blood.

The typical contract was for five years and stipulated that the recruits would work seven-hour days for twenty-four cents a day. When they arrived, they found that they were required to work from dawn to dusk in the boiling sun without any increase in pay. They were not allowed to leave the plantation and were treated with total disrespect. Money was deducted for their food, which was minimal and poor. Lodgings were crowded, uncomfortable and alien to their traditional family and village lifestyles. A substantial percentage got sick and died, compounded by fatigue. These indentured workers were not technically slaves, but the only real difference was they received a pitiful wage.

Appearance of the sugar beet

During the Napoleonic Wars (1803–1815), cane sugar was no longer available in French-controlled Europe due to the naval blockade of the British. This meant that the French sweet tooth had to be satisfied from an alternate source of sugar. Fortunately for them, the solution was already at hand in the form of the humble beet, already a source of food and fodder in Europe (Macinnis 2002; Aronson and Budhos 2010). Back in 1747 Andreas Marggraf at Berlin University had discovered that the beet contained sugar that was indistinguishable from that found in sugarcane. His student Franz Karl Achard developed a process to extract that sugar and built the first sugar extracting factory in modern day Poland. By the nineteenth century, a few sugar beet factories were scattered across France.

In response to the British boycott of sugar, Napoleon ordered the planting of thousands of acres of sugar beets and by 1814 more than 300 factories were making sugar from beets. The amount of sugar in the beets was then much lower than sugarcane and the extraction process was costlier, but it was pretty much the only source of sugar available. When Napoleon's empire collapsed after Waterloo, the boycott was lifted and the cheaper Caribbean cane sugar took back its European dominance.

Sugar from beets remained in the twilight for several decades, until the British banned slavery in the Caribbean, causing cane sugar prices to rise substantially. By now, beets had been selected with sugar levels that were comparable to sugarcane and the cost of extraction had dropped dramatically. Now the two kinds of sugar were on almost equal footing. By 1854, 11% of the world's sugar came from beets and by 1899 65% more sugar was extracted from sugar beets than sugarcane. Today, the tide has shifted to about three-quarters of the world's sugar coming from tropical grown sugarcane and one-quarter from temperate grown sugar beet.

Modern sugar industry

Today, Brazil is by far the leading sugar producer in the world, followed by India, the European Union (EU), China, Thailand and then the United States. Brazil, India and Thailand grow almost exclusively sugarcane, while the EU is predominantly sugar beet, the United States is 55% sugar beet, and China is 20% sugar beet (Huntrods and Koundinya 2012; Anonymous 2014, 2015). Sugar production in the United States and EU are heavily subsidized and imports are tightly regulated. Most of the sugarcane still goes to make sugar, but an increasing amount is used to produce ethanol, particularly in Brazil.

In the sugarcane fields of the world, workers still suffer under hot, brutal conditions and are pushed hard to cut as much cane as possible in the shortest period of time. Many are temporary workers brought in from outside the production regions by labor managers. Often they become almost slaves when money for food, housing and other necessities are deducted from their salary leaving them in debt. In Brazil, the jobs of these miserable workers are being eliminated rapidly by harvesting machines, leaving thousands without their only source of income. Such machines also threaten the jobs of workers in the other cane producing regions.

Worldwide sugar consumption has continued to grow decade by decade, in spite of many artificial sweeteners having been invented such as saccharine (1879), aspartame (1965), Splenda (1976), and the development of high-fructose corn syrup (1976). While the centers of sugar production have continually shifted across the globe over the 10,000-year history of sugar, humankind has not lost its near addiction to the intense, sweet flavor of the white gold with the bittersweet history.

References

Anonymous (2013) *The trans-Atlantic slave trade database*, Emory University, http://www.slavevoyages.org/tast/index.faces.

Anonymous (2014) *FAOSTAT data*, http://www.fao.org/statistics.

Anonymous (2015) *Sugar world markets and trade*, USDA Foreign Agriculture Service, http://www.fas.usda.gov.

Aronson, M. and Budhos, M. (2010) *Sugar changed the world. A story of magic, spice, slavery, freedom and science*, Clarion Books, Houghton Mifflin Harcourt, New York.

Burns, E.B. 1993. *A history of Brazil*. Columbia University Press, New York.

Daniels, J. and Daniels, C. (1993) Sugarcane in prehistory. *Archaeology in Oceania*, vol. 28, 1–7.

Decker, M. (2009) Plants and progress: Rethinking the Islamic agricultural revolution. *Journal of World History*, vol. 20, 187–206.

Dewald, J. (2004) *Europe 1450 to 1789, Encyclopedia of the early modern world*, Charles Scribner's Sons, New York.

Galloway, J.H. (1977) The Mediterranean sugar industry. *Geographical Review*, vol. 67, 177–194.

Gopal, L. (1964) Sugar-making in ancient India. *Journal of the Economic and Social History of the Orient*, vol. 7, 57–72.

Hancock, J.F. (2013) *Plant evolution and the origin of crop species*, Third Edition, CAB International, Wallingford, UK.

Huntrods, D. and Koundinya, V. (2012) *Sugarcane profile*, Agricultural Marketing Research Center, http://www.agmrc.org/commodities__products/grains__oilseeds/sugarcane-profile/#

Klein, H.S. (2010) *The Atlantic slave trade*, Cambridge University Press, Cambridge.

Macinnis, P. (2002) *Bittersweet: The story of sugar*, Crow's Nest, Allen & Unwin, Australia.

Mintz, S.W. (1985) *Sweetness and power. The place of sugar in modern history*, Penguin Books, New York.

Parker, M. (2011) *The sugar barons. Family, corruption, empire, and the West Indies*, Walker, New York.

Sokolow, J.A. (2003) *The great encounter: Native peoples and European settlers in the Americas, 1492–1800*, M.E. Sharp, New York.

Schwartz, B. (1985) *Sugar plantations in the formation of Brazilian society; Bahia, 1550–1835*, Cambridge University Press, Cambridge.

Warner, J.N. (1974) Sugar cane: An indigenous Papuan cultigen. *Ethnology*, vol. 1, 405–411.

Watson, A.M. (1974) The Arab agricultural revolution and its diffusion, 700–1100. *Journal of Economic History*, vol. 34, 8–35.

Williams, M.W. (1930) The treatment of Negro slaves in the Brazilian empire: A comparison with the United States of America. *Journal of Negro History*, vol. 15, 315–336.

2 Banana

Beginnings

The earliest evidence of banana cultivation has been found in the highlands of New Guinea, dating back at least 7,000 years. There in a small area called the Kuk Swamp, twenty feet below the surface, scientists discovered traces of plowing, drainage ditches and even post holes (Denham *et al.* 2003; Koeppel 2009). Shifting through tons of sediment, the team found microscopic phytoliths or stone cells that were identical in shape and size to those found in the bananas of today. All other remnants of the bananas had long since decayed and disappeared, but the durable silica evidence remained.

The first banana species domesticated was *Musa acuminata* spp. *banksii*, which is native to New Guinea (Heslop-Harrison and Schwarzacher 2007; Hancock 2013). It is hard to imagine what humans saw in the wild bananas as they have only a small amount of fleshy pulp that is filled with flinty, teeth-breaking seeds. The botanist Edmond de Langhe has suggested that the banana's starchy corm might have first been harvested and cooked as a vegetable and the consumption of the fruit came later. However, it is not too hard to imagine that those human beings were smart enough to begin farming in the first place could also see the potential in the wild banana. A similar intuitive leap was made by farmers at several distant corners of the world, when they started domesticating an almost lint-free wild cotton, and through years of selection made it into a useful fiber plant.

The transition from wild to really edible bananas must have involved the gradual selection for both seed suppression and parthenocarpy (the production of fruit without seeds). These two transitions had to occur in tandem to reduce seed numbers but still allow the fleshy pulp to develop. Normal fruit development is stimulated by the tiny amounts of hormone produced by seeds. In parthenocarpy, fruit (such as seedless watermelons) can swell and mature without this stimulation, so seeds become unnecessary. The most likely scenario is that the first banana farmers began by transplanting the best wild clones they could find into their gardens. These would have been the plants whose fruit had the most flesh and fewest seeds. Over time these superior clones probably naturally crossed, producing a few superior hybrids that humans recognized and propagated for the next round of planting. It is known that a number of genes exist in wild banana populations for parthenocarpy and seedlessness. These genes are rare and are of no advantage in the wild where seed production is key to long-term survival, but would have been very beneficial in developing a fleshy, seedless fruit for human consumption.

Early spread of bananas

Banana cultivation gradually spread northeast from New Guinea to Indonesia, Malaysia, Borneo, the Philippines and on into southeast Asia, over a period of about 2,000 years (Denham *et al.* 2003; Heslop-Harrison and Schwarzacher 2007; Perriera *et al.* 2011). The first bananas likely arrived in Indian continent somewhere around 4,000–5,000 years ago, as banana phytoliths have been found in 4,000-year-old sediments at Kot Diji, Pakistan.

As banana cultivation spread out from its center of origin, a remarkable amalgamation of cultivated and wild clones occurred. There are a number of different subspecies of *Musa acuminata* scattered across the islands of Micronesia, Malaysia, Borneo and the Philippines. As banana cultivation arrived on each island, the domesticated types likely hybridized with the local subspecies to provide farmers with an ever-expanding choice of hybrids to grow. The only subspecies that did not get incorporated into the banana gene base was *M. burmannica* of mainland southeast Asia.

When banana cultivation reached the mainland, the *Musa acuminata* hybrids began crossing with another wild banana species, *M. balbisiana*. The *balbisiana* types provided increased tolerance to cold and drought, while edibility in *M. balbisiana* was greatly enhanced by crossing with *M. acuminata*. A number of different combinations of *M. balbisiana* and *M. acuminata* arose. If we represent the chromosomes of *M. acuminata* as AA and that of *M. balbisiana* as BB, hybrids appeared that were AB, AAA, AAB, ABB, ABBB and AABB. The hybrid types that are "triploid," with three chromosome sets (AAA, AAB and ABB), were completely sterile and seedless, due to their unbalanced chromosome numbers and irregular assortment of chromosomes during gamete formation.

Varieties representing all the ancient combinations of A and B are still grown, although the most widespread bananas contain three sets of chromosomes. The AAA types, which are often referred to as bananas, produce the sweetest fruit and are eaten fresh; the AAB and ABB hybrids, which are frequently referred to as plantains, are starchier and are cooked or used to make beer.

In a truly remarkable detective effort, molecular biologists have been able to trace the species and subspecies origins of all the modern bananas grown across the world (Perriera *et al.* 2011). They can extract DNA from any banana variety, search for diagnostic fingerprints in the genetic code and trace the origin of the variety's chromosomes to their original south Asian home – much the same as is now done from cheek swabs of humans who want to trace their ancestry. As is described below, this ability to trace a banana's origin has proven to be an important tool in figuring out how bananas were moved across the world.

Early movement of people and bananas in Oceania

Banana cultivation gradually spread from New Guinea to all the scattered islands of near and far Oceania (Greenhill and Gray 2005; Kirch 2010). Surprisingly, it was not the original domesticators of banana that spread it throughout Oceania, but an upstart group, the Lapita cultural complex, who began their island diaspora about 5,500 to 6,000 years ago from Taiwan or southern China. We know that the Lapitas populated the far islands of Oceania, as they left a distinctive trail of pottery behind.

The Lapita migration began about 3,500 years ago and ended about 950 years ago. Their expansion went as far north as Hawaii, as far east as Easter Island, and as far south as New Zealand. They likely even visited the coast of South America, as one of their crops became the sweet potato, a South American domesticate. The Lapita cultivated AAB bananas whose genetic fingerprints tell us came originally from the north coast of New Guinea or the Bismarck Archipelago. This means that they reintroduced AAB bananas to Melanesia, the place where they were originally domesticated.

For decades an argument raged among anthropologists as to whether the Lapita entered Melanesia from southeast Asia, or whether they developed indigenously in New Guinea and the Bismarck Archipelago (Greenhill and Gray 2005). The two models were popularly referred to as the "fast train to Polynesia" versus the "tangled bank." A wealth of linguistic and archeological data has been accumulated over the last thirty years that supports the "fast train" model.

One can't help but be amazed at the courage and confidence it must have taken small groups of Lapita to load up an outrigger canoe with provisions, crops and animals and set off into the vast ocean in search of a new homeland. Their will to establish new outposts must have been strong indeed. James Cook and the other European explorers of Polynesia in the late eighteenth century were stunned and puzzled to find these far-flung islands so well populated by people.

Spread of banana cultivation into Africa

Bananas were first brought into eastern Africa about 5,000 years ago (De Langhe *et al.* 2005; Lejju *et al.* 2006; Neumann and Hildebrand 2009). The molecular fingerprints of modern varieties suggest that AAB plantains were introduced from the area between the Philippines and New Guinea, and AA and AAA bananas from the area between Java and New Guinea.

No one knows for sure who took them there, but banana phytoliths have been found in Uganda in sediments that are 4,700 years old. Over the next millennia or so, banana cultivation spread further west through the tropical middle of Africa. Phytoliths have been found in sediments at ancient agricultural sites in Nkang, Cameroon, that date from 700 BC to 300 BC.

The Bantu peoples likely distributed banana throughout the rest of the rainforest of Africa in their migration from Central Africa across to the African Great Lakes region and down through southern Africa. This migration started about 3,000 years ago at the border of present day Nigeria and Cameroon, and ended a millennium later. As the Bantu civilization began their diaspora, they picked up banana as an important, dependable food source that was well adapted to the tropical rainforest.

Bananas were also introduced to Africa separately from India, 4,000–5,000 years ago. By then, active trading was being conducted across the Indian Ocean via outrigger canoes. Despite the hardships associated with traveling across the Indian Ocean, numerous other crops were moved back and forth between the continents, including sugarcane, coconut, rice, taro and water yam from India to West Africa, and pearl millet, sorghum and cowpeas from Africa to India. Like banana, sugarcane was first domesticated in New Guinea and introduced into India.

In another wave of introductions, the Arabs spread bananas across Northern Africa as part of the crop assemblage they carried in their agricultural revolution. This movement started in Palestine in 650 BC and ended in Moorish Spain about AD 1200. Bananas were also introduced across Africa as a by-product of the slave trade between the Arab countries and North Africa. This sad practice went on from the seventh century AD until the early 1900s.

African bananas now show the most diversity of any bananas in the world, with over 100 varieties grown in the continent. Some of this diversity surely came from the crop being repeatedly introduced from multiple locations, but a significant proportion of the variability probably originated as mutant sports that caught human attention. The banana gene pool would have steadily grown over the millennia, as our clever ancestors discovered more and more desirable mutant sports or shoots and planted them.

Banana to the New World

Banana first arrived in the New World in 1516 when the Friar Monk Tomás de Berlanga planted two rhizomes in his garden in Santo Domingo (Koeppel 2009). He brought the banana from the Canary Islands where it had been introduced by the Spanish after their conquest. The good friar became a tremendous fan of the banana and gave rhizomes to anyone who would plant them. Within a generation the banana was distributed far and wide across the Caribbean, being found wherever Europeans had settled. Berlanga was also responsible for the introduction of the banana to

Central America, taking it there when he became the Bishop of Panama. The banana rapidly became a key food crop in all the southern European colonies, although it took another 350 years before it came to the attention of North Americans. Apples remained the king fruit in the temperate climates until the 1900s.

The first bananas began to arrive in North America in the early 1800s. What were called "Cuban Red Bananas" were sporadically reaching New York and Boston by that time (Raynolds 2003; Soluri 2003). The first regular shipments of bananas into the north began in the 1850s, still mostly red ones from Cuba and Jamaica. The first substantial numbers of yellow bananas made their way to North American ports in the 1870s. These yellow bananas were primarily the variety "Gros Michel," which would come to completely dominate the export industry for the better part of a century.

Gros Michel took a long, slow route to North America and Europe (Soluri 2003; Koeppel 2009). In the early 1800s, the French naturalist Nicolas Baudin collected bananas in southeast Asia and carried them to a botanical garden in Martinique where they flourished. They came to be known as "Figue Baudin" or "Baudin's fig." In 1835, the French botanist Jean François Pouyat carried Baudin's fig to Jamaica, where it was renamed the Poyo banana after its importer. Over the next forty years, this banana variety was spread all across the Caribbean and underwent another name change to Gros Michel. No one knows where this name originated, but as Koeppel (2009, p. 33) remarked in his banana book "the fruit is rather large, and perhaps masculine."

Once Gros Michel found its way to North America, the banana's popularity grew by leaps and bounds. In 1871, the wholesale value of bananas was less than $250,000 (Soluri 2003). By 1901 that value had risen to $6.5 million and in 1911 that value stood at $12.4 million. Bananas were well on their way to being the most popular fruit eaten by North Americans.

Bananas also had become a popular fruit in Europe by the early 1900s. The major suppliers were small-scale farmers in Jamaica and later Belize and Suriname. The European industry was dominated by a single company, Fyffes Limited, which was given a monopoly by the British government. This will be discussed more fully later.

First banana men

The first great banana merchant was Carl Augustus Frank, a German immigrant whose home base was near New York. His infatuation with bananas began in 1865 when he was a steward on a Pacific mail ship (Chapman 2007; Cohen 2012). He saw his first bananas growing along a train track in Colombia and was so fascinated with them that he bought himself a bunch and smuggled them onboard. After an unusually fast return trip to New York of eleven days, his bananas were still green and he sold them for twice what he paid. In 1867 he set himself up as a full-time banana merchant with his brother Otto. They grew their bananas on leased land in Panama and shipped them north in rented space on mail ships.

Lorenzo Dow Baker of Cape Cod became the second great banana merchant (Koeppel 2009; Cohen 2012). He began his career as a sea captain and discovered bananas in Jamaica in 1870 after dropping off a load of prospectors at the Orinoco River, Venezuela. While waiting for his ship to be refurbished in Jamaica for the return trip, he stumbled on bananas and decided to try his luck as importer. Like Frank, he got lucky, made the passage in eleven days, and sold his bananas at the remarkable price of $2 a bunch. This spurred him to buy land and establish a huge estate in Jamaica, where he grew Gros Michael and shipped them north.

In 1885, Baker teamed up with twenty-five-year-old Andrew Preston, a New England produce buyer (Koeppel 2009; Cohen 2012). The two could not have been more different, an old Yankee salt and a refined Bostonian, but their individual strengths were a perfect match. They found eight other investors that were willing to kick in $2,000 each and founded what they called the Boston Fruit Company.

With capital in hand, Preston and Baker abandoned wind-powered ships for ones powered by steam, and began refrigerating their bananas with ice. The bananas now arrived in port in good enough condition to be shipped by train to cities well outside of Boston. They built a series of cold-storage warehouses across the Eastern and Midwestern United States to serve as distribution centers. By the end of the century, the Boston Fruit Company was by far the most dominant banana company in the United States and picked up the nickname of "El Pulpo" ("the Octopus").

Figure 2.1 Captain Lorenzo Baker, who first brought Jamaican bananas to the United States

Source: US Library of Congress.

The largest rival to El Pulpo was Standard Fruit, a company established by Joseph Vaccaro and his brothers in New Orleans. They shipped their bananas out of Honduras and supplied them to the southern United States. The Vaccaro brothers built their niche in the banana business by buying virtually all the ice factories ringing the Gulf Coast. This gave them a stranglehold on southern refrigerated transport.

The Central American expansions of Minor Cooper Keith

By the late 1800s, the Caribbean could no longer produce enough bananas to satisfy the North American markets. To meet this burgeoning demand, the banana merchants began to focus their attention on Central America, where vast areas were being deforested and planted into bananas. At the epicenter of this activity was Minor Cooper Keith in Puerto Rico (Chapman 2007; Koeppel 2009; Cohen 2012). He and his two brothers were developing huge plantations there as they carved a railroad from the capital of San José to the eastern port of Limón.

The railroad was a monumental task, as it cut through barely penetrable stretches of land covered with swamps and dense forest. Temperatures skyrocketed each day and the air was teeming with clouds of mosquitos; the dew point was dripping and rain fell regularly, turning the cleared earth into muddy mire. The project was supposed to take three years but wound up taking sixteen years, suffering through multiple periods of cash and manpower shortages.

The brutal conditions made finding workers extremely difficult. They started with native Costa Rican workers who very quickly had had enough, requiring the recruitment of an ever-changing cast of workers – first Jamaicans, then Italian immigrants from the United States and finally prisoners from New Orleans jails. Many of these workers were so horrified when they saw what was in store for them that they disappeared into the forest, never to be heard from again. It is estimated that 5,000 men lost their lives on the project, dying from malaria, yellow fever, dysentery and dehydration. Even Keith's two younger brothers succumbed to malaria.

The planting of bananas by the Keiths began as a modest effort to feed the workers before growing into a giant export business. The enterprise was fueled by sweet land deals negotiated with local dictators. The first of these deals came in 1882, when the railroad project stalled about thirty miles from Limón due to insufficient funds. Keith was able to convince the Costa Rican President Guardia to loan him a million dollars from the country's treasury to keep the project going. In addition, he obtained a ninety-nine-year concession to run the railroad, full control of the port at Limón and 800,000 acres of land adjacent to the railroad tracks (tax free). The first "Banana Republic" was born.

Keith continued to build railroads all across Latin America and receive greatly favorable land deals everywhere he went. He gained control of huge acreages in Panama, Guatemala, Honduras, Nicaragua, Colombia and Ecuador. He developed plantations in all these countries and gained control of a vast shipping network, ultimately becoming the major exporter of bananas to the American southeast. By his mid-30s, Keith had become the richest man in Central America and by far the most powerful.

Keith managed to go broke again in 1899, and this time it was Preston of Boston Fruit that came to his rescue. Preston covered his debts and the two companies were merged together to form the giant company, United Fruit. Preston became president and Keith vice president. United Fruit immediately set out to root out all their competitors. Some of these they bought out, but many were forced into bankruptcy by United Fruit by reducing their prices to ridiculously low levels. At its peak, United Fruit controlled 80%–90% of the bananas that were brought and distributed in North America.

The last great banana man – Samuel Zemurray

The last of the great banana merchants was Samuel Zemurray, who came to be called "Sam the Banana Man" (Chapman 2007; Koeppel 2009; Cohen 2012). Zemurray was a Russian immigrant who came to the United States in his teens and worked his way up from stacking shelves in his uncle's dry goods store to junk dealer, fruit peddler and ultimately banana tycoon. He got his start in the banana business in Mobile, Alabama, when he bought a few thousand bananas that had been dumped to the side of the wharf by the Boston Fruit Company. They had deemed the bananas too ripe for shipping and sold them to Zemurray for a little over $100 (almost all the money he had in the world). Zemurray rented a box car and traveled with his bananas to Selma, selling them at each stop to local grocers and peddlers. He managed to unload most of the bananas before they became overripe, and made a profit of about $40.

From this modest beginning, Zemurray gradually expanded his shipping network across the south. In 1899, he moved only about 20,000 bananas, but by 1903 that number had risen to over a half a million. He became so successful that he entered into a contract with United Fruit to become their primary shipper. However, Sam did not want to be just a merchant; he wanted the whole pie. He found a partner named Hubbard and they bought an aging Steamship Company to bring his own fruit from Central America; United Fruit became a 25% silent partner in this venture.

Hubbard and Zemurray then purchased the Cuyamel Fruit Company in Honduras, whose land holdings Zemurray quickly expanded from 100 to 5,000 acres, using borrowed money. This venture scared off Hubbard, but gutsy Sam now had himself a whole banana company – producing, shipping and selling. He then went about maximizing his profits by convincing the president of Nicaragua, Miguel Dávila, to make Cuyamel Fruit exempt from taxes and duties. This deal ultimately led to a revolution, as will be described later.

Don't cross Sam

At the same time that Zemurray was building his banana business in Nicaragua, the country was facing a monumental debt crisis (Koeppel 2009; Cohen 2012). Nicaragua owed British creditors over $100 million and the Brits were out of patience. The creditors had begun pleading with the British

Navy to intercede. The original loan was taken out in 1860 to finance the building of a railroad that went bust and had been left unpaid for forty years, compounding interest. The threat of British intervention in Nicaragua raised the concern of the US president, William Howard Taft, who through the Monroe Doctrine could not let that happen. He instructed his secretary of state, Philander Knox to find a solution.

Knox approached the fabled financier J. Pierpont Morgan to refinance the debt and get it out of the hands of the British. Morgan agreed to take on the debt, but only if the Morgan Bank could have representatives in the custom house of Nicaragua's main port, Puerto Cortés, to collect a duty on all imports and take a cut before passing on the rest to the government. He further stipulated that this plan had to be ratified as a treaty by the Nicaraguan Assembly. Tightly squeezed President Dávila acceded.

The backlash to this plan became significant. Not only did the people of Nicaragua consider it an encroachment on their national sovereignty, but Zemurray was particularly enraged that it would eliminate his sweet deals with the country and add a penny to the cost of each bundle of bananas. He decided to initiate a publicity campaign, hiring lobbyists to convince US congressman to pressure the White House and supporting writers of newspaper articles and editorials. The overriding theme Zemurray concocted was that the United States should not meddle in the internal affairs of other countries. This was, of course, counter to his own mode of operation!

The campaign was so successful that Zemurray was called into the secretary of state's office for a dressing-down. This further enraged Zemurray and he decided that the path he must take was to foster a revolution to overthrow President Dávila. Zemurray, a thirty-two-year-old Russian immigrant, had decided to defy the US president and secretary of state and wrest control from a sovereign nation!

To lead the "revolution," Zemurray picked General Manuel Bonilla, who he had gotten to know as an expatriate living in New Orleans and a disposed president of Nicaragua. To lead the army of liberation, Lee Christmas was selected by Zemurray, a soldier of fortune who had fought alongside Bonilla in Nicaragua. Christmas put together a motley army of about 100 characters from the bordellos and bars of New Orleans and other gulf ports. Zemurray secretly bought a 160-foot warship to get them to Nicaragua, and within a matter of weeks Christmas's band of misfits took over the country largely unmolested, picking up bands of rebels along the way. Dávila abdicated after a long negotiation on a US gunboat, passing rule to an interim president until new elections could be held. US Secretary of State Knox was forced to reverse himself and support the deal.

Bonilla won the election in a landslide. Not surprisingly, his first act was to sign a bill allowing Zemurray to operate in the country tax free. He was granted the right to import materials and equipment duty free, and was allowed to build anything anywhere to support his business (railroads, highways or buildings). He was also given a loan of $500,000 to repay him for the cost of the revolution and 24,700 acres of land for future expansion. Zemurray now virtually owned a country.

Life of banana laborers

Banana workers in Latin America were paid much higher wages than those toiling on sugar and coffee plantations, but similar to their compatriots they were treated almost as slaves (Forster 2003; Chapman 2007; Koeppel 2009). The company completely controlled their lives – deciding when and where they worked and firing them at will for even the most minor of infractions. The banana laborers were paid in scrip that could only be redeemed by purchasing goods at the company store, which charged outrageous prices. The workers had no contracts and were "on call," only being paid when there was a job to do. The field managers acted almost as overseers, many being from the southern United States, carrying fond memories of slavery before the Civil War.

The worst job of all was clearing the forest, work so horrendous that a continuous stream of Afro-Caribbean laborers had to be recruited, as the local people would have nothing to do with it. These workers lived in tents or porous palm-thatched huts with no protection against the clouds of ravenous mosquitos. The temperature and humidity skyrocketed each day. They labored from 4:00 a.m. to about noon, when the heat made further strenuous work impossible. Each worker was expected to clear a hectare of forest in a six-day workweek. An ever-present danger was the venomous snake, *barba amarilla*, two and a half meters long with no warning rattle. Scorpions dropped off of trees and repeatedly stung the sweating, exhausted workers.

Once the jungle was cleared, irrigation ditches were dug and plants were set in rows on mounds. The plants grew to ten feet within a year and produced their first crop. The ripening of groves was timed so that bananas were produced continuously on a plantation. Gangs roamed the fields hacking at weeds with machetes, keeping irrigation ditches clear of debris, pruning plants to a single cropping stem and applying fungicides and pesticides. Those who applied chemicals to control disease turned blue from the copper sulfate used, eventually losing their appetites and then dying.

When the green bunches of fruit were ready to harvest, one man would climb up each plant and cut the stem free, dropping the fruit to another man who hoisted it upon his back and jogged to a railroad car. These men worked at short notice, being summoned into action after a ship had called ahead, alerting the company of their pending arrival. No women were allowed in the field as the carriers had to have great physical strength. They made dozens of trips each day hauling bunches of fruit weighing up to 100 pounds. Back injuries and ugly knife wounds were common as fatigue grew and attention waned; twelve-hour days were the norm. Once the loaded train got to port, sirens blared, calling stevedores to work, on a first-come, first-hired basis. They worked for up to ten hours a day hauling the heavy bunches of fruit back and forth between train and ship until the hold was completely loaded.

The harshness of the day laborer's life on a banana farm cannot be better described than by Cindy Forster in her article "The Macondo of Guatemala":

> Work as a company plantation laborer tested the limits of human endurance, with long days working through torrential rains that cut visibility to several feet and churned the ground to a sea of mud, and heatwaves so intense they transformed the coastal terra caliente into a pressure cooker. It was a regimen that devoured workers.
>
> (2003, p. 203)

Worker unrest and revolt

As the twentieth century dawned, banana workers became increasingly restive about their brutal work conditions (Chapman 2007; Koeppel 2009). They were increasingly aware of the true value of the crop they handled and how little of that value came back to them. They also were appalled at the amount of land that was under control of the banana companies, who didn't think twice about expelling indigenous people. Thus began a long tortuous history of violent labor unrest and bloody reprisals by the banana companies, local dictators and even the US military. Worker demands revolved around their being treated with dignity – fair wages, reasonable work hours, good medical care and clean toilet facilities. They wanted to be paid in cash, not scrip; to have the necessities of life priced fairly; and to be hired as salaried employees, not as temporary subcontractors.

A major outbreak of violence occurred in Guatemala in 1910, when 600 Jamaicans rioted against company prices and looted two United Fruit stores. When the government refused the company's pleas to intervene, a group of local Guatemalans (*ladinos*) took matters into their own hands and lynched a Jamaican. In a tepid response, the police eventually arrested a single ringleader, but they let him go a few days later. United Fruit later rewarded this man by hiring him as a farm superintendent.

This riot was followed by a strike of stevedores in Guatemala, who were enraged by a change in the way they were paid. Instead of the existing flat rate of $1.25 for a ten-hour day, the company decided they would pay a piece rate of twenty-five cents per 100 stems carried. Someone in an office back home had deemed the workers could carry 700 stems a day and actually make more money. In reality, few stevedores could manage even half that amount and would effectively be taking a huge pay cut. United Fruit asked the local dictator, Manuel Estrada Cabrera, for military help to no avail, and eventually were forced to return to the old day rate.

The Guatemalan government did respond a few months later to another plea from the company, when Jamaican laborers rioted in revenge after a ladino murdered one of them in a dispute over cards. Troops arrived in the night and fired randomly into the Jamaicans sleeping quarters, murdering dozens. A few months later, United Fruit took it upon themselves to break a strike in Boco del Toro, Panama. They made arrests, had ringleaders killed and even burned down worker dormitories and uprooted the plants in their gardens.

Jamaican field laborers struck in Colombia in 1918, demanding an eight-hour day, a six-day workweek and better health care – the same work conditions that their white managers received. United Fruit turned a deaf ear, this time letting the fruit rot in the field until the workers got hungry enough to come back. It was easy for the company to choose this strategy to break the strike, as banana demand was in a slump after World War I, and they couldn't have sold most of the fruit anyway.

Throughout the 1920s, labor unrest spread to all of the Republics of Central America. The founding of the Soviet Union in 1917 had inspired worker movements all across the world and the US government became obsessed with limiting the spread of communism in their hemisphere. US Marines were repeatedly sent to Central America to quell banana strikes and restrict any strikes or political changes deemed unfavorable to US interests. In 1920, Guatemalan farm workers tried to unionize, but their revolt failed when US Marines were sent by President Franklin Roosevelt to "protect" US interests. Workers in Panama went on strike in 1925, only to have US troops intervene again. In reality, nothing could be done in Central America without US consent; US soldiers occupied Nicaragua continuously from 1912 to 1933, the Dominican Republic from 1916 to 1924, and Honduras from 1911 to 1925.

The wave of insurrection among banana workers reached a crescendo in late 1928 and early 1929 when 32,000 workers went on strike in the Magdalena region of Colombia. Vigilante groups were organized by United Fruit and trained by the US military to try and quell the walkout. Martial law was declared. Sadly, it ended when multitudes of strikers and their families gathered in the town square of Ciénaga after Mass, waiting for a speech by a regional governor. The general in charge, Cortés Vargas, gave the crowd five minutes to disperse and when they didn't, he had machine gunners on rooftops fire on the crowd, killing a thousand people. The jittery general's explanation was that he had had the choice of quelling the riot himself or the United States would have intervened again. This action spiraled into the decades of guerilla warfare and bloody battles that still plague Colombia.

Great White Fleet

Beginning in the late 1800s, a vast network of refrigerated cargo ships was making regular trips from Caribbean ports to the United States and Europe (Chapman 2007; Koeppel 2009). Painted white to reflect the sun, they were named "La Gran Flota Blanca" (the Great White Fleet) by Captain

Lorenzo Baker. The Great White Fleet was the largest private navy in the world.

As time went on, United Fruit built and obtained larger and larger cargo vessels. They chartered and reconfigured four new ships from the US Navy that were not finished in time for the Spanish-American War. In 1911, they built the massive ship, the S.S. *Sixaola*, which was half as big as the *Titanic* and could hold 500 railcars of bananas. By this time, even the smallest ships of United Fruit could carry thousands more bananas than Baker's original fleet.

The Great White Fleet played an important role in World Wars I and II. Many of the ships were used to transport foodstuffs, military vehicles and troops across the ocean. Unfortunately, the big, ponderous ships served as easy targets for roaming German U-boats. Overall, nineteen United Fruit ships were sunk during World War II.

The S.S. *Sixaola* transported tens of millions of tons of beef during World War I, until it caught fire and partially capsized in New York in February of 1917. It was rebuilt and went back to service as a cargo vessel for a couple of decades until being called back into action in World War II. It was sunk in June 1942, when two torpedoes ripped into its hull near Guatemala and its boiler exploded. Twenty-nine of the crew died in the attack and forty-nine drifted aimlessly in the ocean for four days until they hit dry land.

War between Cuyamel and United Fruit

In 1915, an almost open war erupted between Cuyamel Fruit and United Fruit (Cohen 2012). Zemurray had had the audacity to obtain land along the Utila River in the Motagua Valley, previously the sole domain of United Fruit. He obtained land on the Guatemalan side of the River as part of dictator Bonilla's promise of 24,700 acres and he bought land on the Nicaraguan shore by outmaneuvering United Fruit. Zemurray paid both of two disputing "legal" owners for the land, while United Fruit was wasting time and money on lawyers trying to identify the true owner. The CEO of United Fruit, Victor Cutter, cried bloody murder claiming that Zemurray had violated an international border, but to no avail. Cuyamel Fruit forged ahead clearing land and planting on both sides of the river.

The two companies began a campaign of dirty tricks, their agents crossing over the river at night and raising havoc – slicing water lines, smashing equipment and turning over trucks and trains. Zemurray began smuggling guns and ammunition to Honduran rebels in the hills. The Port Authority in New York discovered a Cuyamel boat filled with weapons, and officials in New Orleans found another cache of weapons in a house. United Fruit bought land in the highlands of Nicaragua, to control the flow of the rivers that irrigated Cuyamel land. When Zemurray asked for permission to build a bridge between his properties on each side of the river, he was denied by Honduran President Tiburcio Carías, now in the pay of United Fruit.

Zemurray responded by building two long wharfs on either side of the river and an inflatable "dock" that could be deployed between them in three hours.

The battle between the two companies escalated into a major international incident, with the governments of Guatemala and Honduras now drawn into the fray. By 1928 the tensions had reached the point that the US State Department decided they had to defuse the situation. The only logical solution was to convince the two parties into a buyout or merger. These actions would defy existing antitrust legislation in the United States, but the alternative was deemed worse. The combined companies would own 68% of the world's banana business.

Zemurray and Cutter were called to Washington, DC, to negotiate a deal. After days and days of tedious negotiations, Zemurray finally agreed to a merger. The final deal was reached in a restaurant over twelve pints of beer by Zemurray and Bradley Palmer, United Fruit's largest shareholder. Cuyamel would become a division of United Fruit and stock would be swapped. Zemurray received 300,000 shares of United Fruit stock, making him one of the richest men in the United States. Not bad for a Russian-born immigrant who arrived in the United States almost penniless. As part of the deal, Zemurray had to retire and get out of the banana business. To keep himself occupied, he went into a frenzy of building and philanthropy. He built a plantation, golf course, hunting lodge and game park. He gave boat loads of money to Tulane University, Radcliffe, and Harvard. He built a clinic for troubled children and a hospital for black women.

Don't underestimate Sam

During Zemurray's retirement, the profitability of United Fruit took a huge tumble (Cohen 2012). The profits of the company fell from $45 million in 1928 to only $6 million in 1932. Stock that was worth $158 a share in 1930 was valued only at $10 a share in 1932. The stock market crash of 1929 played a significant role in the loss of revenue, but the major contributing factor was that the company was no longer run by a banana man. Cutter had died and management had been transferred to the major shareholder, Daniel Gould Wing, the president of the First National Bank of Boston. None of the major stockholders, except Zemurray, had direct knowledge of the growing and transportation of bananas.

At the board meeting in 1932, Zemurray expressed his concern and offered suggestions for improvement. He had visited the operations in Nicaragua and thus had ground level information. When the Board decided to deny a request for $10,000 from a Guatemalan manager to build an irrigation ditch, Zemurray implored them to either trust their local manager or fire him. He offered numerous suggestions on how the company's operations could be made more profitable. Wing and the board ignored all his propositions.

Enraged, Zemurray decided he would gather proxies from other shareholders and take control of the company. He would be "the fish that ate the whale," as Rich Cohen (2012, pp. 142–143) titled his book on Zemurray. Visiting shareholder after shareholder, he gradually gathered enough support to take over. At the next board meeting in January 1933, Zemurray again made his suggestions and again was rebuffed. He was cut off by Wing who rudely stated, "Unfortunately, Mr. Zemurray, I can't understand a word of what you are saying." Zemurray stormed out of the meeting, gathered his proxies, returned and fired Wing. He ended the meeting with the words, "You gentleman have been fucking up this business long enough. I'm going to straighten it out." Zemurray had been banned from starting a new banana business, but not from taking control of United Fruit.

Zemurray Heads Company; Coolidge Named Chairman

—Photo by Blank & Stoller.
SAMUEL ZEMURRAY

ZEMURRAY CHOSEN FOR PRESIDENCY OF UNITED FRUIT

Managing Director Succeeds to Post of Late F. R. Hart

Samuel Zemurray of New Orleans was elected president of the United Fruit Company Monday afternoon at a meeting of the board of directors in Boston, Mass. Mr. Zemurray, who for the past five years has been managing director of the company, succeeds Francis R. Hart, who died on January 18.

At the same time T. Jefferson Coolidge, a great-grandson of President Thomas Jefferson, and an undersecretary of the treasury during the first term of President Roosevelt, was elected chairman of the board of directors.

Others Named

Other officers named were Arthur

Figure 2.2 Zemurray takes over United Fruit

Source: February 1, 1938, edition of the *Times-Picayune*.

Banana marketing routes

Beginning at about the turn of the century, the worldwide banana trade had become dichotomous (Raynolds 2003). Most of the bananas produced in Latin America went to the United States and the bulk of those grown in the Caribbean traveled to Britain. This separation of trade routes came in response to British policies supporting their colony of Jamaica and later their Windward Island colonies. To prevent a United Fruit monopoly, the British government gave the company now known as Fyffes 75% of the British market and sole rights to those bananas produced in Jamaica and later Belize and Suriname. They made a similar deal with the UK food company, Geest Corporation, for the Windward Islands. These arrangements were kept in force when the British Caribbean colonies gained their independence in the 1950s and 1960s.

The two banana marketing routes were supplied by two very different farming structures. The bananas going to the EU markets were produced by thousands of small-scale farmers in the Caribbean, who intercropped their bananas with other fruit and agronomic crops on farms generally less than ten acres. Their fruit was brokered by state-sponsored associations. The producers in Latin America were the vertically integrated, export companies themselves, growing their bananas in monoculture on tens of thousands of acres.

The two systems of banana farming had very different agro-ecological and social implications. The huge monocultures of Latin America required copious applications of pesticides, sickening and sterilizing field workers and causing cancer and birth defects in children of the women in the packing sheds. The more complex intercropping systems of the Caribbean required far less pesticide application and in many cases the growers could not afford them at all. As a result, many fields in the Caribbean were pesticide free and aerial spraying was simply not used. Because the acreages were so much smaller, there was considerably less deforestation in the Caribbean, resulting in far less soil erosion and siltation. The farmers in the Caribbean had to work long, brutally hard hours, but unlike the farm workers in Latin America, the profits gained from their labor were their own.

The banana republic

By the turn of the twentieth century, Guatemala was essentially owned by United Fruit (Schlesinger and Kinzer 1990; Chapman 2007; Koeppel 2009). At least half of the Guatemalan economy was bananas and United Fruit had 75% of that business. At its peak, United Fruit owned four million acres of the country, which was 70% of the arable land. Most of that land was uncultivated forest, being held in reserve for future expansion as United Fruit fled from land infested with Panama disease.

Over a span of forty-five years, a series of Guatemalan dictators were more than happy to allow United Fruit full run of the country in return for financial backing, a stable economy and infrastructure development. The first of these despots was Manuel Estrada Cabrera who began his rule in 1898, allowing no freedom of the press and keeping power by murdering anyone who opposed him. When he was finally deposed in 1920, a revolving door of five dictators followed over the next decade, all ruthless and all married to United Fruit.

General Jorge Ubico took power in 1931 and held it for over a decade. Like all his predecessors, he ruled with an iron fist, being particularly harsh on the native Mayan peoples of Guatemala. He required them to work for landowners at least 100 days a year. He maintained a secret police force and declared that any Indian workers who didn't follow orders on a farm could be summarily executed. Ubico was obsessed with communism and outlawed the use of the words – "trade union," "strike," "petition" and even "worker." He also was a great friend of United Fruit, allowing them to undervalue their vast landholdings and pay no taxes.

By 1944, the people of Guatemala had had enough of Ubico's repressive regime and began protesting in the streets. Ubico tried to halt the strikes by having his soldiers fire indiscriminately into the crowds, but the protests continued and grew until the outrage reached such a crescendo that Ubico was forced to resign. He tried to hold on to power by putting his loyal crony General Juan Federico Ponce in charge. This move failed when two junior officers, Captain Jacobo Arbenz and Major Francisco Arana, orchestrated a military takeover and all the top generals were killed. Arbenz and Arana called for free elections, and in 1945 a university professor named Juan José Arévalo took charge of the country.

Then came what Guatemalans called "Los Diez Años de la Primavera" ("Ten Years of Springtime"). During his term in office, Arévalo allowed political parties, gave the press its freedom, started social welfare programs and even allowed workers outside of the banana plantations to form unions. He left the sweetheart deals with United Fruit intact, preferring to hold the giant at bay. Arévalo miraculously survived numerous assassination attempts and coups and finished his term of office in 1950. He was succeeded by Jacobo Arbenz, one of the men responsible for Ubico's overthrow.

Arbenz became the hated enemy of United Fruit. He was appalled by how much of Guatemala was owned by the company and became a staunch supporter of land reform. He originally came from the Indian highlands where he had learned firsthand about the dismal living conditions of the native Mayans. He wanted to make Guatemala much more independent economically, and in his inaugural address he railed against private ownership of farms. His old partner Francisco Arana, now the military leader of Guatemala, did not share his zeal for reform and was mysteriously murdered.

The giant growls

Soon after Arbenz took office, officials of United Fruit visited him to demand that their special relationships with Guatemala remain unchanged (Schlesinger and Kinzer 1990; Chapman 2007; Koeppel 2009). Arbenz was not receptive and suggested instead that export duties would be imposed and that all new land acquisitions would be priced according to real value. The United Fruit officials left in a huff, appalled that they would be required to share a piece of their massive profits.

Zemurray and United Fruit decided that they had no alternative but to get Arbenz removed from office, using whatever dirty tricks were necessary to achieve that goal. They began by hiring the famous public relations giant Edward Bernays, whose assignment was to persuade America that Arbenz was a communist and an enemy of the United States. He took journalists (on United Fruit's ticket) to Guatemala on "fact-finding missions," which resulted in dozens of anti-Arbenz exposés being published in leading newspapers and magazines. United Fruit produced a slick documentary called "Journey to Banana Land," which showed scenes of happy workers and described all the wonderful things United Fruit had done for Guatemala. The movie became a staple of US classrooms.

Arbenz held his ground and in 1952 made his "Decree 900." In it, the government would take over all unplanted, private farms that were larger than 223 acres and redistribute the land to peasants. The original landowners were to be compensated for the stated value of their land. This action resulted in over a quarter of a million acres being transferred to 100,000 families.

Decree 900 hit United Fruit particularly hard, as they were holding most of those unplanted acres. Adding salt to the wound, they were given only $600,000 in compensation – the value reported on their own tax returns! They howled in protest, and even got the US State Department to demand payment of $15 million, but the Arbenz government turned a deaf ear.

Enraged Zemurray decided that they had no alternative but to start another publicity campaign, this one being directed to US government officials. Arbenz would be sold as a communist, in a country now embroiled in the throngs of anti-communist McCarthyism. They hired the newspaperman John Clements to produce a "report" that linked Arbenz to the Soviet Union and claimed he was planning to take control of the Panama Canal. This largely fanciful "Report on Guatemala" was sent to 800 governmental legislators and aides on the right.

The US government was predisposed to accept Clements's propaganda, as many of its high-ranking officials had strong ties to United Fruit. The Secretary of State, John Foster Dulles, had been a partner in the New York law firm of United Fruit. His brother Allen, the head of the CIA, was a former board member of United Fruit. United Fruit's head public relations man, Ed Whitman, was married to President Eisenhower's personal secretary.

Ugly American takeover

To prove his administration was tough on communism, President Eisenhower decided that Arbenz had to go and ask the CIA to do the job (Schlesinger and Kinzer 1990; Chapman 2007; Koeppel 2009). A team was soon assembled with E. Howard Hunt at its helm. The plan was to select and train a small invasion force and use psychological warfare to shatter the resolve of Arbenz and his military. United Fruit would pay the bills. Colonel Carlos Castillo Armas was chosen to led the revolutionary army. He had been a part of a previous plot to overthrow Arbenz and was readily available, now selling furniture in Honduras. A group of 400 soldiers was recruited in the United States and sent to Honduras for training by the CIA.

Arbenz got wind of the plan and was convinced his forces would be crushed unless he could buy sufficient weapons to arm them. He tried Denmark, Mexico, Argentina and Switzerland before finally obtaining a boatload of old German World War II weapons from Czechoslovakia. The CIA found out about the deal and allowed the boat to dock in Guatemala, offering its arrival as proof of Arbenz's communist ties.

The "revolution" began with a blockade of the ports of Guatemala by the US Navy. Arbenz appealed to the Organization of American States and the United Nations to no avail. US planes began to fly low and harass the capital and military bases, dropping propaganda leaflets all across the country. United Fruit beamed pro-revolution propaganda over the airwaves from

Miami, calling it "La Voz de la Revolución" ("the voice of the revolution"). Fake reports of Arbenz committing suicide were circulated and published by local papers.

The invasion itself turned out to be a dismal failure. In the first confrontation in Zacapa, 122 revolutionaries were routed by thirty Guatemalan soldiers, with only twenty-two surviving. A rebel force of 170 was defeated by dock workers when they tried to take the major port of United Fruit, Puerto Barrios. The survivors of these battles limped back to Nicaragua.

However, the airways and rumor mill told a different story to the world and an increasingly agitated Arbenz. Reports circulated that thousands of Guatemalans were joining the revolution, and thousands more were seeking refuge in Nicaragua. There were fake accounts of pitched battles that the revolutionaries won, complete with sounds of rifles, bombs and screams. Huge numbers of Guatemalan soldiers were said to be switching sides. To keep US public engaged, the US ambassador fed them fake photographs of widespread atrocities being conducted by Guatemalan soldiers.

Arbenz became convinced that Eisenhower would soon send American soldiers and initiate an all-out invasion. There were even rumors that US soldiers had already landed on shore. Depressed, obsessed and confused, Arbenz asked his military leaders to arm the peasants in a last-ditch effort to save the country. Almost paralyzed with fear, they refused.

Arbenz roamed the halls of his palace for seven more sleepless, drunken days until he decided to resign on June 27, 1954, and get out of the country to save his life. Without the military's support, he felt he could no longer go on. In a final indignity, he was stripped to his underwear and searched at the airport before fleeing to Mexico with his family. Cameras clicked and the images were shared across the world. Mighty United Fruit had control of the country back.

Demise of Big Mike

Virtually all the bananas being grown in the early to mid-nineteenth century in Latin America and the Caribbean were clones of an original, single progenitor of Gros Michel (Koeppel 2009; Cohen 2012). Thus, almost all of the plantation bananas had identical genetic backgrounds, and any disease that killed one of them would eventually kill them all. This was a disaster waiting to happen.

The first major disease threat to Gros Michel found its way to Panama from southeast Asia in the early 1920s. What came to be called Panama disease was a soil-borne pathogen that attacked roots. The leaves of infected plants became withered and the whole plant eventually toppled over. Gros Michel had no resistance to the disease and an effective control measure proved elusive. The only solution was to continually start new plantations on virgin ground, making sure to get planting stock from the remaining clean plantations.

Once Panama disease struck a plantation, it was abandoned, leaving ghost towns behind. Anything reusable was removed and transferred. United Fruit became obsessed with owning as much non-cultivated forest as possible to hold in reserve. By 1940, it owned 50% of the private land in Honduras, but had only 10% under cultivation. Standard Fruit also expanded its landholdings, but not nearly as dramatically, as they were a much smaller company.

The second grave threat to Gros Michel came in the form of another disease, Black Sigatoka. The leaves of infected bananas turned brown and crinkly as if they had been burned, reducing yields by more than 50%. Sigatoka was caused by an airborne spore that was easily moved from field to field on muddy boots and equipment. First seen in Fiji in 1911, it reached Colombia in 1930 and began a steady march through Central America, hopping relentlessly from plantation to plantation. By 1931, over 50,000 acres in Panama were infected.

United Fruit scientists did find a way to control this disease, but it had nasty ramifications. They discovered that the Bordeaux mixture used to control rots in grapes could also control Sigatoka, if it was sprayed repeatedly on the plants in massive quantities. They delivered it by plane and overhead irrigation as many as fifty times a year. The fruit, plants, ground and cities turned blue, being stained by the active ingredient, copper sulfate. The men charged with its application also turned blue, lost their appetites and wasted away. They were called "Los Pericos" (the parrots), enticed by high wages to do this nasty work.

While the banana companies could control Sigatoka through great effort, they had no solution to Panama disease except to keep moving. United Fruit was convinced in the 1950s that flooding fields could hold the disease at bay, but they were pretty much just fooling themselves. The only real solution would come from finding a variety that was resistant to the disease. United Fruit did little on this front, as they had seemingly limitless land for expansion, and feared that the American public would reject any banana but Gros Michel. Standard Fruit, with much more limited acreage, had to find a resistant variety or expire.

Welcome Cavendish

Standard Fruit began searching in earnest for a resistant banana variety in 1927 (Koeppel 2009; Cohen 2012). They experimented for a while with the Lakatan banana from the Philippines, but it proved to have a multitude of problems. It was a much darker banana than Gros Michel, was sensitive to temperature and humidity, and individual fruit periodically dropped off bunches. The biggest problem was that Lakatan fruit didn't ripen regularly and the bunches had to be exposed to ethylene gas to induce proper ripening. Ethylene is the same hormone that naturally triggers ripening in apples and other fruits. To produce ripe Lakatan bananas for the retailer, controlled atmosphere storage rooms would have to be built at US distribution centers. The marketers had little interest in doing this as long as United Fruit was shipping plenty of "Big Mike."

Standard Fruit also experimented with Dwarf Cavendish, a close relative of the Cavendish variety, which would ultimately become the banana of choice. The fruit of this variety were rather small, but they ripened more regularly than Lakatan and grew better. It was also resistant to Panama disease and high winds. Unfortunately, it never took off, as it was very thin-skinned and had to be handled with great care.

Standard Fruit became very excited in the 1940s with a banana hybrid called IC2, developed at the Imperial College of Tropical Agriculture in Trinidad. Breeders there had filtered through tens of thousands of cross-pollinated bananas to retrieve a handful of seeds, the winning one being a hybrid between Gros Michel and a wild Asian banana. IC2 proved to be an excellent shipper and initially was resistant to both Sigatoka and Panama disease. About a half a million bunches of IC2 wound up being shipped to the United States in the late 1940s; however, its popularity proved ephemeral. It lost its resistance to Panama disease and proved to be very difficult to grow on all but the very best soils.

Another banana Standard Fruit tried was Bout Rond, introduced from Puerto Rico. It had excellent ripening characteristics and even looked like Gros Michel. However, it was never widely planted, as it turned out to be extremely susceptible to Sigatoka, bruised and rotted easily during transport and readily toppled over in high winds.

Standard Fruit finally found their replacement for Gros Michel in Cavendish. No one knows for sure where it originally came from or what it was first named, but it is known that in 1826 Charles Edward Telfair had plants of it in his botanical garden on the African island of Mauritius. It is possible that he got his cuttings from China, either through his extensive plant trading network or an indentured Chinese worker who came to work the sugar fields. Telfair subsequently sent cuttings of this plant to England, where it was eventually passed on to what became its namesake, William Cavendish, the Duke of Devonshire. From the duke's estate, it began a worldwide trot, finding its way by 1850 to almost all places where bananas were grown, except Central America. Its journey became complete in 1939, when it was introduced into Honduras from Brazil.

Cavendish was resistant to Panama disease. It was still susceptible to Sigatoka, but it was shorter in stature and more easily drenched in Bordeaux mixture. Its fruit tasted great and the plant stayed upright in hurricane winds. The only major problem with Cavendish was that its fruit were much more fragile than Gros Michel and couldn't simply be thrown into the holds of ships without severely bruising them. Standard Fruit solved this problem by putting the fruit in boxes, revolutionizing the way bananas were handled, shipped and marketed. The fruit could now be transported by trucks, spelling the end of railroad transit.

Standard Fruit sent the first shipment of Cavendish to the United States in 1947, and by 1958 all their banana cargos were Cavendish. Stodgy United Fruit finally gave up and began shipping Cavendish in 1960, completely copying Standard Fruit's handling techniques. The last Gros Michels arrived in the United States in 1965. The banana industry had been completely transformed.

The difficult world of banana breeding

For over 150 years, only two banana varieties, Gros Michel and Cavendish, have been widely grown in Latin America and the Caribbean. With most other crops, new replacement varieties are developed regularly by breeders and a single variety rarely dominates in any one place for long. The predominance of Gros Michel and Cavendish has been to a large part due to the reluctance of the banana companies to challenge consumer preferences, but also because of the genetic difficulties associated with banana breeding (Simmonds 1995; Koeppel 2009; Ortiz 2013; Guilford 2014).

Remember that most bananas are sterile and produce very few seeds, if any. Cavendish produces no seeds, and Gros Michel generates only one or two in thousands of fruits. Other varieties and wild selections produce more seeds than these two varieties, but in general a breeder must make 1,000 hand pollinations on 200 plants to get enough seeds to find a single promising selection. A quarter of a million fruit must be sorted through to get a couple dozen seeds – a daunting task to say the least.

The first breeding work on banana was begun in 1922 at the Imperial College of Tropical Agriculture established by the British government. Working with a vast array of germplasm collected from across the world they did manage to release a hybrid resistant to Panama disease, but it proved less than desirable for many other characteristics. United Fruit opened a research station in Honduras in 1925 and hired the great USDA germplasm explorer, Wilson Popenoe, to establish an extensive banana collection there, but most of his work focused on crop and disease management. Popenoe led the effort to develop the Bordeaux mixture to control Sigatoka.

Numerous additional banana breeding programs have popped up across the world, but none has yet produced a replacement for Cavendish. The breeder who has come closest is Phil Rowe, whose work was first privately funded by United Fruit and later publicly funded as the Fundación Hondureña de Investigación (FHIA). From 1959 until his suicide in 2001, Rowe generated 40,000 hybrids and released over twenty banana varieties. Most of these are grown locally somewhere in the world, but do not have the proper array of characters to topple Cavendish. His most popular variety, called the "Apple banana" or "Goldfinger," has big, firm fruit, never turns brown, and is resistant to Panama disease and Sigatoka, but has heavy flesh and is too tart for American tastes.

Rony Swennen at the International Institute of Tropical Agriculture in Nigeria has also been a highly successful banana breeder, beginning his work in 1979. His work has not focused on dessert bananas but instead on the cooking bananas or plantains of Africa. By crossing fertile plantain varieties

Figure 2.3 Banana infected with the new, devastating Race 4 of Panama disease

Source: http://www.promusa.org/Fusarium+wilt.

with infertile ones he has produced a number of tasty and Sigatoka-resistant varieties that will be important hedges against hunger in subsistence agriculture. His research group also developed the first genetically modified (GM) bananas in the early 1990s.

Banana breeders are currently scrambling to find resistance to two major disease threats. A new strain of Panama disease (Tropical Race 4) has emerged in Asia that all banana varieties including Cavendish are highly susceptible. The disease has now spread all across southeast Asia and just recently jumped to Jordan and Mozambique. If Race 4 follows the same course that Race 1 did in the first half of the twentieth century, it is only a matter of time until all the bananas in Latin America and the Caribbean are ravaged. So far, no resistance has been identified to this pathogen.

Another disease called Banana Xanthomonas Wilt also severely threatens small subsistence farmers across Africa. It is the most devastating disease on the continent, causing 80%–100% yield loss. To date, no varieties are known to be resistant, although GM bananas have been developed carrying resistance genes from sweet pepper (Namuddu 2014). These GM types are currently in confined field trials to test their efficacy and environmental safety.

Recurring disease threats to the banana industry are not going to go away until the companies stop the cultivation of a single variety. With a wider diversity of varieties being grown, either conventionally or biotech derived, it is much more likely that at least a few will be resistant to any new strain or disease that emerges. As N.W. Simmonds (1995, p. 374) put it, "In the longer run, the ultra-conservative trades will probably have to change their habits and grow new clones if they are to survive at all; they used to say that they could not grow Cavendish clones!"

Company flip-flops and subsequent tragedy

During the 1960s, the face of the Latin American and Caribbean banana industry began to shift dramatically (Chapman 2007; Koeppel 2009). Standard Fruit was purchased by Castle & Cook, a pineapple importer, and renamed Dole. Del Monte got into bananas by buying the West Indies Fruit Company and taking control of a number of United Fruit farms in Guatemala. United Fruit became heavily branded as "Chiquita," with a sexy, swinging banana as its logo. A Wall Street takeover artist named Eli Black acquired United Fruit and merged it with his own company into United Brands. All the banana companies began to diversify into other fruits, foods and snacks. They began to sell their farms in Central America, becoming fruit buyers, leaving the headaches of fruit growers to local interests.

Profitability of the banana business slid precipitously in the early 1970s, particularly for Chiquita. A more potent strain of Sigatoka emerged in Honduras requiring more aerial spraying, increasing the costs of banana production and further endangering the health of ground workers. In a federal antitrust suit, Chiquita was forced to sell huge holdings in Guatemala to Del Monte. As their profits slid, they started a price war with its former subsidiary Fyffes, eventually chasing them out of Central America. Accusations raged that United Fruit had instigated the destruction of Fyffes banana shipments.

Chiquita lost $2 million in 1971, but Black was convinced that he could bring Chiquita back. He sold land, diversified and received many kudos by pouring money to rebuild Nicaragua after a devastating earthquake. Chiquita's profits rose to $16 million in 1972, but by 1974 the company was again losing millions of dollars. Black just couldn't get a break. Costa Rica, Panama and Honduras banded together to form the "Organization of Banana Exporting Nations" and began charging $1 a box on exported fruit. Panamanian workers went on strike. Sigatoka continued to rage and a hurricane wiped out three-fourths of Nicaragua's crop. For the first time, Chiquita fell behind Dole in US sales, ending a dominance of close to 100 years.

On Feburary 1, 1975, Black took the elevator up to his office on the 44th floor, smashed the window with his briefcase, fastidiously cleared up the shards of glass on the floor and hurled himself through the window. It is clear that he had had enough. Not only could he not stop Chiquita's financial bleeding, but it was later discovered that he had gone against his own religious beliefs and bribed the president of Honduras with $1.25 million for tax relief.

Banana trade wars

By the early 1990s, the European banana market had become highly desirable, now exceeding that of the United States. To bolster its efforts to protect the African, Caribbean and Pacific banana industry ("ACP bananas") from the Central and South American corporations Dole, Chiquita and Del Monte ("dollar bananas"), European governments initiated tariffs in 1993. The argument was that dollar bananas could be produced much more cheaply than ACP bananas, and the only way that the ACP banana industry could survive was to protect it. Otherwise, tens of thousands of small scale farmers would lose their livelihood (Raynolds 2003, Anonymous 2009, Koeppel 2009).

Chiquita screamed bloody murder. They now had 20% of the European banana market and would lose at least half of that, costing the United States what they estimated to be $400 million in sales. They were forced to divert more of their bananas to the United States, driving prices and profits down dramatically. Chiquita approached the Clinton administration in 1996 and got it to lodge a complaint with the World Trade Organization (WTO) declaring unfair and illegal trade practices. To seal the deal, the current owner of Chiquita, Carl Linder, donated $500,000 to the Democratic Party after meeting with the Clinton administration.

In 1997, the WTO ruled that the EU tariffs did violate international free trade agreements. The EU made a counterproposal that was rejected, beginning a period of over ten years where the issue went back and forth between the EU and the US government (really Chiquita). Along the way, ten Latin American countries joined the fray. Finally, in 2009, a compromise was reached where the EU agreed to reduce tariffs gradually over a seven-year period, from €176 to €114 per ton. The hope was that the lower tariff would keep the ACP growers at least competitive, without giving them a prohibitive advantage. To help bolster the ACP countries, the EU provided a compensation package of almost €200 million over four years. Only time will tell if these tariffs and compensation packages can keep the small farmers of the ACP countries afloat.

Banana industry today

The volume of global banana exports continues to rise, primarily from Latin America and the Caribbean (Anonymous 2014). The United States and Europe are by far the largest banana importers, sharing evenly about 65% of the world's production. Ecuador is now the largest banana exporter in the world, followed by Costa Rica, Colombia, and Guatemala. Overall South American production is currently declining slightly compared to Central American production. Exports from the Caribbean to Europe have shown a slight recent decrease, with a shift toward more local sales.

The worldwide banana market remains a battleground of a few monster US and British corporations, supported strongly by their home governments. Probably gone are the days of active military intervention by the US government into the affairs of Central America, but well-placed sums of money and cronyism still have considerable sway. The huge, negative environmental impact of the monoculturing of a single variety has not gone away, resulting in massive deforestation, environmental poisoning and human illness. Workers are still badly mistreated and unions are periodically crushed depending on the attitudes of ever-changing governments. Promising breeding work is now being conducted at a number of locations that could diminish the threat of disease on banana productivity, but the marketers and American public will need to accept more than one variety to prevent periodic disease catastrophes. It is fair to say that fewer men have total power over the banana industry today, and Central American countries are left more frequently to their own devices, but corporate policies still rule the day.

References

Anonymous (2009) Q & A: The banana wars, *BBC News*, http://news.bbc.co.uk/go/pr/fr/-/2/hi/business/8390099.stm.

Anonymous (2014) *FAOSTAT data*, http://www.fao.org/statistics.

Chapman, P. (2007) *Bananas: How the United Fruit Company shaped the world*, Canongate, Edinburgh.

Cohen, R. (2012) *The fish that ate the whale: The life and times of America's banana king*, Picador, New York.

De Langhe, E., Pillay, M., Tenkouano, A. and Swennen, R. (2005) Integrating morphological and molecular taxonomy in *Musa*: The African plantains (*Musa* spp. AAB group). *Plant Systematics and Evolution*, vol. 255, 225–236.

Denham, T.P., Haberle, S.G., Lentfer, C., Fullagar, R., Field, J., Therin, M., Porch, N. and Winsborough, B. (2003) Origins of agriculture at Kuk Swamp in the highlands of New Guinea. *Science*, vol. 301, 189–193.

Forster, C. (2003) "The Macondo of Guatemala" Banana workers and national revolution in Tiquisate, 1944–1954. In: *Banana wars: Power, production and history in the Americas* (pp. 191–208). Striffler, S. and Moberg, M. (eds.). Duke University Press, Durham and London.

Greenhill, S.J. and Gray, R.D. (2005) Testing population dispersal hypotheses: Pacific settlement, phylogenetic trees, and Austronesian languages. In: *The evolution of cultural diversity: Phylogenetic approaches* (pp. 31–52). Mace, R., Holden, C. and Shennan, S. (eds.). UCL Press, University College, London.

Guilford, G. (2014) How the global banana industry is killing the world's favorite fruit, *Quartz*, http://qz.com/164029/tropical-race-4-global-banana-industry-is-killing-the-worlds-favorite-fruit.

Hancock, J.F. (2013) *Plant evolution and the origin of plant species*, Third Edition, CAB International, Wallingford, UK.

Heslop-Harrison, J.S. and Schwarzacher, T. (2007) Domestication, genomics and the future for bananas. *Annals of Botany*, vol. 100, 1073–1084.

Kirch, P. (2010) Peopling of the Pacific: A holistic anthropological perspective. *Annual Review of Anthropology*, vol. 39, 131–148.

Koeppel, D. (2009) *Banana: The fate of the fruit that changed the world*, Plume, Penguin Group, New York.

Lejju, J.B., Robertshaw, P. and Taylor, D. (2006) Africa's earliest bananas? *Journal of Archaeological Science*, vol. 33, 102–113.

Namuddu, A. (2014) *Genetically modified crops under research and development in Africa*, http://www.nepadbiosafety.net/subjects/biotechnology/gm-crops-under-research-in-africa.

Neumann, K. and Hildebrand, E. (2009) Early bananas in Africa: The state of the art. *Ethnobotany Research & Applications*, vol. 7, 353–362.

Ortiz, R. (2013) Conventional banana and plantain breeding. Proceedings of the international ISHS-ProMusa symposium on bananas and plantains: Towards sustainable global production and improved uses. *Acta Horticulturae*, vol. 986, 177–194.

Perriera, X., De Langheb, E., Donohuec, M., Lentferd, C., Vrydaghse, L., Bakrya, F. and 12 others (2011) Multidisciplinary perspectives on banana (*Musa* spp.) domestication. *Proceedings of the National Academy of Sciences*, vol. 108, 11311–11318.

Raynolds, L.T. (2003) The global banana trade. In: *Banana wars: Power, production and history in the Americas* (pp. 23–47). Striffler, S. and Moberg, M. (eds.). Duke University Press, Durham and London.

Schlesinger, S. and Kinzer, S. (1990) *Bitter fruit: The story of the American coup in Guatemala*, Harvard University Press, Cambridge, MA.
Simmonds, N.W. (1995) Bananas: *Musa* (Musaceae). In: *Evolution of crop plants* (pp. 370–375). Smart, J. and Simmonds, N.W. (eds.). Longman Scientific and Technical, Harlow, UK.
Soluri, J. (2003) Banana cultures: Linking the production and consumption of export bananas (pp. 48–79). In: *Banana wars: Power, production and history in the Americas*. Striffler, S. and Moberg, M. (eds.), Duke University Press, Durham and London.

3 Cotton

The curious evolution of the cotton species

There are about forty-two wild species of cotton found in four continents of the world, but only four species were domesticated:

- The tree cotton (*Gossypium arboreum*), now grown to a limited extent in India and Pakistan;
- Levant cotton (*G. herbaceum*) with a current, spotty distribution in Africa and Asia;
- Sea Island or extra-long staple cotton (ELS) (*G. barbadense*) grown mostly in Egypt;
- Upland or Mexican cotton (*G. hirsutum*), now domesticated all over the globe and representing 95% of the total cotton production.

Unraveling the evolution of the domesticated cotton species has not proven to be an easy task (Wendel 1989; Wendel *et al.* 1989; Brubacker *et al.* 1999). First of all, Levant cotton has a known, wild ancestor, but tree cotton does not. Tree cotton was long thought to be a derivative of Levant cotton, but the two species are so different genetically that it is more likely that they were domesticated from different species. No one knows who that other progenitor might be, as nothing like it now exists in the wild. The ancestor must be extinct.

Second, Sea Island and upland cotton have twice as many chromosomes as tree and Levant cotton, and that extra chromosome set came from a rather improbable source. Based on a wide array of genetic data, it appears that the chromosomes of Sea Island and upland cotton came from two ancestral species, South American *Gossypium raimondii* and African *G. herbaceum*. The problem here is, of course, that these two ancestors are located on different continents. Strong evidence has shown that Sea Island and upland cotton emerged long after Gondwanaland separated into the continents of Africa and South America.

Two explanations have been given for how the two progenitors came together: (1) some unknown people carried an Old World species to

America and the hybridization occurred, or (2) the direct ancestor of Sea Island and upland cotton arose in South America after a seed of a relative of *Gossypium herbaceum* drifted across the ocean from Africa or Asia and germinated.

Most cotton researchers favor the idea that the seed was naturally dispersed rather than carried by humans and that they arrived in the Americas across the Pacific, rather than the Atlantic. This makes the most sense, as the native *Gossypium raimondii* are located on the western side of the New World. A primitive progenitor of Sea Island and upland cotton probably arose in northwestern Mexico and then evolved into two separate species as it spread into Peru. Genetic evidence suggests that *G. herbaceum* is the mother (seed parent) of that initial hybrid and *G. raimondii* the father.

Cotton domestication and evolutionary change

Remarkably, the four cotton species now grown were domesticated about the same time (5,500 to 6,000 years ago) on four different corners of the earth (Hutchinson 1951; Brubacker *et al.* 1999). In the New World, Sea Island cotton was likely domesticated in central-coastal Peru and upland cotton in the Tehuacán Valley of Mesoamerica (Stephens 1976a). In the Old World, humans first domesticated tree cotton in the ancient agricultural village of Mehrgarh in the Indus Valley. Clear evidence of the first domestication of Levant cotton is much more elusive than the other cottons, but seeds from its wild progenitor were likely collected in eastern Africa by Indian traders and moved north into cultivation.

It must have taken a real leap of faith for people to begin cultivating cotton, as the wild populations show little promise of generating sufficient quantities of fiber for textile production. The wild species produce sparse amounts of coarse seed hairs that are short (about one millimeter long) and would be very difficult to spin into thread. The wild plants are also gangly, perennial shrubs and their seeds germinate with difficulty. What may have given people a clue about cotton's potential was the fact that they were already making textiles from wool and flax fibers, and were well versed in spinning and weaving. Still, a lot had to happen to the cotton plant before it could be used to make cloth.

Most believe that the cotton plant became what it is today after people began to cultivate it (Hutchinson *et al.* 1947; Stephens 1967; Wendel *et al.* 2010). The ancient farmers must have imagined cotton's potential and gradually improved it by painstakingly selecting seed from the most promising plants generation after generation. The great cotton geneticist S. G. Stephens called this process "evolution under domestication."

The first priority of ancient peoples must have been to improve the quality and quantity of the seed hairs to make them more spinnable. Seeds of our cultivated cotton are covered by two types of fibers, short fuzz hairs (much like the seed hairs of wild plants) and long, spinnable lint hairs that are absent on seeds of wild plants. People would have had to start by first identifying mutants with a few spindly lint hairs and then start selecting for increased quantities of them. This likely took many generations.

As the quantity of lint increased in proportion to fuzz, people would have focused more and more on fiber length, strength and color. Also, types with more permeable seeds would have been selected that germinated more readily and uniformly. This later trait might have been selected for unconsciously by humans when they scattered seed for planting. Plants from the fastest germinating seeds would be more likely to reach maturity and be harvested than plants from later germinating seed.

Another road block limiting the utility of wild cotton for textiles would have been the difficulty in separating the lint from seeds. Our ancestors must have placed strong selection pressure on the proportion of lint versus fuzz, and ease of lint removal (called ginning). Cotton scientists have pointed to two races, wild species that would have provided the most easily ginned seed, at least in the New World – the kidney-seed forms of *Gossypium barbadense* and the tufted-seed forms of *G. hirsutum*. In the tufted-seed forms, the lint is concentrated on one end of the seeds, making its removal easier; in the kidney-seed types, the seeds are all fused at one end of the pod, which allows the seeds to be removed in clumps, rather than individually.

As domesticated cotton spread across the Old World, it was moved from its native tropical and subtropical environments into more temperate conditions (Stephens 1976b). To allow for this dispersal into cooler environments, early maturing, day-length neutral types would have been selected that could be harvested before frosts. In essence, humans had to change the native cotton plant from a perennial that produced bolls in the winter to an annual that could be harvested in the summer. These "day-neutral" cotton types may have first been selected by humans inadvertently in tropical climates as they attempted to find shorter season types that could be harvested before late season droughts.

Cotton's dramatic transformation under domestication is a testament to the ingenuity of ancient peoples. Through foresight and diligence, humans in several far flung societies changed a marginally useful species into what became their primary source of textiles. As Brubacker *et. al.* (1999) state: "human selection transformed sprawling perennial shrubs or small trees with impermeable seeds bearing sparse, coarse hairs to compact annualized plants with large, readily germinable seeds bearing a thick coat of long, strong, white fibers, the cotton of commerce." They further observed that "the cotton species now exhibit far greater levels of morphological diversity and encompass much greater geographic ranges than their antecedents." The remarkable change in cotton from its humble beginnings is a clear example of the power of human selection.

Old cotton system in India

For thousands of years, India was the center of cotton production in the Old World (Schlingloff 1974; Yafa 2005; Riello 2010). Cotton was grown primarily on small farms as a companion to rice and other foodstuffs. Different types of cotton were grown in different regions of India depending on local climates and soils. These varieties produced varying fiber qualities and this led to specialization across the subcontinent, with low-priced textiles being produced in western India and higher-priced ones along the eastern coast. Fabrics became known by the region producing them, such as calico from the southwestern port of Calicut. In fact, many of the words that we now use to describe fabrics are Indian in derivation, including chintz, gingham, khaki, madras, muslin and seersucker.

Cleaning and spinning the raw cotton was done in the peasant household, largely by women. The cotton bolls were first cleaned of plant trash and stones and then the seeds were removed by a process now called ginning. This was first done by plucking out the seeds by hand, but this excruciatingly slow process was replaced by a machine composed of two rollers and a crank. The cotton fiber was pulled through the rollers leaving the seeds behind. Once the seeds were removed, the cotton was then batted with a bow to remove the remaining dust and dirt. The vibration of the string also loosened the fibers so that the cotton could be twisted into rolls and spun into thread. This was done initially on simple, handheld spindles, but spinning wheels were in common use by the fourteenth century, probably arriving in India from the Middle East.

Men were largely responsible for weaving the yarn into cloth. Whole households were devoted to weaving; one or two men and their helpers wove the cloth on looms and sold their product to the highest bidder. Horizontal looms were used for simple fabrics, while vertical looms were employed to produce textiles and carpets with complex designs. In many villages, half of the households were weavers.

Complex mercantile systems arose where the cloth passed through several hands before finally reaching an export merchant. Often individual weavers would receive advances from a merchant to carry them through the season, as the production of cotton textiles was a long and drawn-out process. To produce a piece of cloth three yards long and one yard wide, it took about eight days – two days to clean and gin the raw cotton, five days to spin the yarn and one day to actually weave the fabric.

Cotton textiles were decorated by complex processes including dyeing, painting and printing. Regional specializations in how the cotton was adorned spread out across India. In Gujarat and the Malabar Coast in west India, what came to be called chintzes were printed with wooden blocks (chit was the Indian word for printing). On the Coromandel Coast in southeastern India, painting was used to produce calico in a complex process requiring numerous dye transfers. As time went by, the names calico and chintz came to be used interchangeably, but initially calico represented unglazed fabric with small geometric or floral designs, while chintz was glazed and shiny, with dramatically colored flowers, birds and animals.

Indian cotton becomes a global commodity

The ancient Indus Valley civilizations first started to trade cotton fabrics to Mesopotamia about 2,500 years ago and to Europe a few centuries later (May and Lege 1999; Yafa 2005; Riello 2010). The Greeks and Romans imported cotton fabrics, although they were considered a luxury item; most of their cloth was made of flax, hemp and wool. Likewise, the ancient Egyptians favored linen, but still imported some cotton as a novelty. Alexander the Great played an important role in spreading knowledge of cotton across the civilized world. When he arrived in Pakistan in 325 BC, cotton was the fiber used to make the loincloths and long shirts of the common folk. His men returned home marveling about what the historian Herodotus called "wool bearing trees with fruits of downy fleece."

By the first century AD, cotton was being traded from India across the Indian Ocean to far-flung ports in Arabia, Africa and Malaysia. The Arabs in the seventh century were carrying what they called *qutun* or *kutun* in caravans across overland trading routes through Asia. The "Silk Roads" that passed through China to central Asia and the Middle East became cotton roads in the reverse direction. Some cotton even found its way on the overland routes to Venice, the primary European center of Asian imports, but it remained largely ignored until the 1600s in England, France, Germany and northern Europe, where wool remained king.

Trade boomed across the Indian Ocean for centuries, and cotton textiles became a part of ordinary people's lives across a broad swath of coastal Asia and Africa. Cotton clothes and decorations were being traded to northern and western Africa for ivory, gold and slaves. Melaka in Indonesia became a particularly important center of trade, where cotton changed hands and continued en route to central Asia and China. Rice, fish, sugar and cotton textiles were traded for spices, sandalwood, porcelain and silk. In the 1500s, when the first Europeans arrived in the Indian Ocean via the Cape of Good Hope, they were astonished at the breadth and complexity of the trade going on in this part of the world. The reach of cotton had at this point far exceeded that of wool, linen and silk.

Early spread of Old World cotton cultivation

Cotton cultivation first began to spread into the Persian Gulf and North Africa about 2,500 years ago, and on into China and southeast Asia about 1,000 years ago (Hutchinson 1951; May and Lege 1999; Riello 2010; Brite and Marston 2013). Cotton was an important part of the Arab agricultural revolution that enveloped Persia, Arabia, Anatolia, northern Africa and ultimately southern Spain. Tree and Levant cotton pretty much traveled the same paths east across the Old World, although only tree cotton took hold in ancient China and southeast Asia.

Cotton cultivation spread rather slowly across the world due to the sophistication needed to process cotton's fibers. It was one thing to plant and grow cotton, but it was of no use, except as an ornamental, until people knew how to spin and weave its fibers. People had known for centuries how to process wool and flax fibers into cloth, but the shorter cotton fibers were much more difficult to spin and weave.

By the eleventh century, cotton cultivation and manufacturing were well established in southern Arabia, the Mesopotamian steppe and the Levant. China had a booming cotton textile industry by 1250. Under Ottoman rule, cotton production further expanded into Anatolia and Egypt in the fifteenth and sixteenth centuries. The cotton textiles produced in the Ottoman Empire were sold mostly within its borders, although many found their way into eastern and western Europe. Almost all the cotton produced in China stayed in the country.

Even though the area of cotton cultivation steadily expanded during the Middle Ages, India kept its stranglehold on the global textile market. As late as the mid-eighteenth century, Indian cotton manufacturing was responsible for almost 25% of the world's industrial output. This is particularly astonishing as the whole industry was based on small weaver households. India held onto the lion's share of the global market because of their well-established trade routes and most importantly, the unrivaled quality of their finished product. For centuries, their competitors could only imitate.

Europe discovers cotton in the New World

When Columbus arrived in the New World in 1492, he found the native Arawaks to be quite adept in cotton growing and textile manufacturing (Phillips 1979; Berdan 1987). Columbus took this as a strong sign that he was on the coast of India, where he knew that cotton textiles were made. The early Spanish historian Peter Martyr, who chronicled Columbus's voyages, made many references to cotton growing, spinning and weaving, and he described how cotton was used to make fishing nets, as well as ornamental wall hangings and clothes. Columbus himself commented in his journals on the scanty cotton clothes worn by the Arawaks: "Here we saw cotton cloth, and perceived the people more decent, the women wearing a slight covering of cotton over their nudities."

On his last voyage to the New World, still searching for a pathway to India, Columbus stumbled on a building in Cuba with thousands of pounds of cotton, along with spindles and looms. In his travels, he reached as far as the Yucatan Peninsula where he found the native Mayans adorned with cotton breechcloths and sleeveless shirts. Columbus was greatly impressed with their creative, sophisticated weaves and vibrant colors.

When the Spanish explorers arrived in Mexico and South America in the fifteenth and sixteenth centuries, they also discovered that the Aztecs and Incas were producing cotton textiles. Cotton fabrics were considered luxury items, with nobles adorning themselves with richly colored, embroidered mantels, shirts and skirts. These early explorers found a prehistoric cotton belt stretching from the southwestern United States through Mexico, Central and South America to the Caribbean. Remarkably, almost the same spinning and weaving technologies were being used across this vast area as in the Old World centers of production.

Perhaps Stephen Yafa (2005) in his book "Cotton: The Biography of a Revolutionary Fiber" summarized the global parallels in cotton production best:

> At approximately the same time in human history, 5,500 years ago, civilizations living as many as 10,000 miles apart and totally isolated from one another domesticated the plant and converted it to fabric. Each invented nearly identical tools – combs, bows, hand spindles and primitive looms – to clean the fiber, then spin and weave it. That remarkable parallel discovery seems to confirm that inspired insights leap past all geographical boundaries to become the common property of mankind. (p. 13)

The Europeans fall in love with chintz

What many historians have called a "chintz craze" began in Europe about 1600 and steadily grew in strength, reaching its peak at the end of the century (Parakunnel 1924; May and Lege 1999; Yafa 2005; Riello 2010). The fine, printed textiles came almost exclusively from India; the cottons from the New World would not have an impact in Europe for a couple of centuries.

Originally chintzes were used as wall coverings, pillowcases, curtains, bed covers and rugs, but by 1800 they were widely used for clothing by people from all walks of life. Some even converted what were originally meant to be bed covers, drapery and rugs into skirts, blouses and gowns. As Stephen Yafa (2005, pp. 31–32) quotes Daniel Defoe: "the chintz was advanced from lying upon their floors to their backs, from their footcloth to the petticoat." The poorer classes were actually the first to wear the cheap but colorful chintzes, but the rich soon followed. A satirical writer suggested that "it became difficult for the better folk to know their wives from their chambermaids."

So why did the Europeans fall in love with the Indian cotton textiles? First, they were cheaper and could be washed much more easily than wool and silk. The cotton fabrics maintained their color much better after repeated washings and when exposed to bright light. Perhaps most importantly, Europeans fell "head over heels" for chintzes because of their unique, bold and flamboyant designs.

Many of the designs flooding into Europe were truly breathtaking. As Stephen Yafa (2005) describes:

> Flower bulbs, jumping fish, birds, butterflies, and sprays of blossoming buds coil about delicately meandering branches that spread out from the sinuous tree trunk, luminescent reds, blues and yellows and a host of other vibrant colors spring to life in designs that are so fresh and buoyant after 300 or more years they appear to be newly created.
>
> (p. 29)

As the chintz craze gathered steam, the local textile industries began to howl in protest. Literally hundreds of pamphlets decried that the widespread sale of Indian textiles was unpatriotic, was produced by pagans, encouraged increased spending by the poor, hurt national wealth, and put already struggling artisans out of work. Daniel Defoe also wrote: "About half the woolen manufacture has been entirely lost, half of the people scattered and ruined, and this by the intercourse of the East India trade."

France ultimately banned cotton importation of all kind in 1686; England began with a partial ban in 1702, followed by a full ban in 1721. However, these laws had little real impact, except to encourage the development of a domestic textile printing industry. The laws were flagrantly ignored by the people infatuated with chintz and the bans were little enforced by the authorities. Except for a few unfortunate women being stripped naked by mobs of angry textile workers, the chintz craze continued largely unabated.

When the bans on cotton import were finally lifted in the mid-1700s, a full-fledged cotton textile printing industry existed in Europe. English and French printers were unashamedly copying Indian printed designs using imported raw materials from India and the West Indies. The early European attempts at printing textiles were poorly colored and faded easily, but the processes of Indian artisans gradually crept into Europe, first from immigrant Armenians who had been shipping calicos all across the Ottoman Empire from about 1600. There were also some attempts to bring Indian artisans directly to Europe, but these efforts proved futile due to extreme cultural differences. Several Frenchman went to India to study techniques of dyeing, printing and painting and published lengthy reports. A particularly infamous one came from the French Jesuit priest, Father Coeurdoux, who was stationed at a mission on the Coromandel Coast between 1742 and 1747. He betrayed the trust of the Indian dyers in his flock by revealing their secrets in a series of published letters.

The textile printing industry that arose in Europe was quite different from that of India. Over a hundred people could be employed at a single factory, a far cry from the simple weaver households. Labor was divided into specialties such as designers, drawers, colorists, painters and printers. In the push for faster production, numerous innovations emerged including copper versus wooden printing plates and ultimately rotary printing. A team of workers who once produced only a handful of printed pieces a day could now generate hundreds.

The booming European textile printing industry of the mid to late 1700s came as a heavy blow to the Indian weaver households. Thousands were put out of work as European textiles became increasingly cheaper and plentiful. The home-based industry of India, whose products had dominated the global market for thousands of years, was losing its grip. This was just a herald of things to come, as the whole Indian textile industry would collapse in the mid-1800s when the great textile mills of England began to spew out cheap cloth.

Emergence of British cotton manufacturing

As the demand for cotton textiles exploded in Europe, it was only a matter of time until a booming domestic manufacturing industry emerged. What started as a cottage industry became by the 1770s an Industrial Revolution (Riello 2010). British ingenuity led the way, stimulated by the ever growing thirst for more cotton yarn and finer textiles.

The first major technological innovation came from John Kay in 1733. He invented a mechanical loom with a "flying shuttle" which was hurled back and forth, weaving the horizontal "weft" yarn through the vertical "warp" threads. Before this innovation, women had to stand on each side of the loom and slowly thread the weft through the warp. The flying shuttle made weaving twice as fast and greatly reduced labor need. It also began the mechanization of the weaving industry, a harbinger of the Industrial Revolution.

Unfortunately, Kay made very little money from his invention. Other textile manufactures stole his design, and while Kay filed numerous lawsuits for infringement, the penalties received were far less than his legal costs. Adding to his woes, mobs of traditional weavers smashed his looms and threatened his life. He fled to France to start anew, but alas, he was no more successful there. The real profits were made by others who shamelessly copied his design.

As the faster, more efficient looms became widespread, so did the demand for cotton yarn. The shortage became so critical that the London Society for the Encouragement of Arts, Manufactures and Commerce offered a reward to the person who could significantly increase the production of yarn. The man who rose to the occasion was James Hargreaves in 1764, who invented the "spinning Jenny" (named after his daughter). This innovation allowed a single woman to spin yarn onto multiple spindles and match the output of eight to ten women using traditional spinning wheels.

As with the flying shuttle, the spinning Jenny incited riots of traditional spinners, forcing Hargreaves to move his operations to another city. He did manage to make a little money from his factory, but not from his innovation, as it too was widely copied without royalty payments. Hargreaves received a grant of £5,000 from Parliament for his invention.

While the spinning Jenny revolutionized the rate by which yarn could be produced, the product was only strong enough for weft and not warp. This meant that linen or wool had to be used as warp by the English weavers. The resulting cloth, called "fustian," was inferior to the much softer, all cotton textiles of the Indian artisans and the dominance of Indian textiles remained.

Richard Arkwright and John Ray (not related to the inventor of the flying shuttle) were the first to develop a mechanical spinning machine that produced cotton thread strong enough for warp. They perfected a process that stretched and twisted the raw cotton during the spinning process, resulting in a much stronger thread than the spinning Jenny. It built on Hargreaves ideas and borrowed liberally from another machine developed by John Wyatt and Louis Paul in 1738. Their innovation was powered by running water and came to be known as the "water frame."

Arkwright became the only British textile innovator that made any real money off his invention, liberally stealing ideas and ruthlessly dropping investors, partners and collaborators along the way. He was also the man who started the Industrial Revolution, a story that will be told later.

The next major innovation in cotton textile production came from Samuel Crompton in 1779. Crompton, who was a weaver himself, invented the "spinning mule," which produced thread faster and of a higher quality than anyone thought possible. It got the name "mule" because it combined components of the spinning Jenny and the water frame. The spinning mule could produce a pound of yarn in twelve hours versus several days with traditional machines. Most importantly, its quality finally rivaled the Indian made threads.

Unfortunately, Crompton was the latest in the long line of textile innovators who made little money from their insight. Instead of patenting the spinning mule, he tried to make a profit by building and then licensing machines. This yielded him little cash, as others found it more economical to simply copy his design.

Through the seventeenth century, the energy used to run the textile machines passed through a number of transitions from human crank, to horse drawn, then water driven and finally steam powered. The great textile plants of the mid-1700s were fueled by water power, but they were replaced by steam power when Cartwright invented the "power loom" in 1785. He was the first to put James Watt's 1769 invention of the steam engine into large-scale manufacturing.

Like many of his predecessors, Edmund Cartwright had to contend with enraged hand weavers, who burned down one of his early, steam-powered mills. This event encouraged him to get out of the mill-building business. Parliament awarded him a meager grant of £10,000 for his innovation, which he parlayed into a farm that supported him the rest of his life.

The Industrial Revolution begins

Richard Arkwright had a vision. He saw a huge manufacturing empire based on his water frame and related technologies that he could beg, borrow or steal (O'Brien *et al.* 1991; Riello 2010). He built his first factory in Cromford, a sleepy little village with an important resource – a rushing river to turn his water wheels. Cromford had long been the home to many traditional spinners and weavers of wool; this was to dramatically change.

He built his first mill in 1771 on a scale of astonishing proportions – a towering brick building with five floors and a huge water wheel to power his spinners. Over a hundred employees worked around the clock to produce cotton yarn that was shipped to weavers all over England. Within a decade he had expanded to twelve factories and almost 1,000 employees, and by 1790 he had 200 factories and 5,000 workers. Arkwright held a virtual monopoly on cotton yarn for two decades, importing his raw cotton from far-off India and the West Indies.

Because of his mills' remote locations, Arkwright had to draw employees from hither and yon. Because his machines did most of the complex work, he needed mostly unskilled workers, which included children as young as eight years old. He employed whole families, and nearly two-thirds of his employees were children. He built row houses to serve as dormitories and churches to serve their spiritual needs. Men, women and children worked a thirteen-hour day, from 6:00 am to 7:00 p.m., with a half-hour breakfast break and forty minutes for dinner. He gave his employees a holiday week each year, but didn't allow them to leave the factory premises.

Figure 3.1 Richard Arkwright, textile tycoon and father of the Industrial Revolution

Source: Gallery of Portraits, volume V (1835), via wikimedia.org.

In the decades of Arkwright's expansions, the spinning of wool essentially disappeared across England. Large buildings of all descriptions were fitted into loom shops. Between 1775 and 1783, calico production vaulted from 56,000 to three million yards in England. Raw cotton imports rose from 4.7 million pounds in 1771 to 56 million in 1800.

By the 1800s, Edmund Cartwright's power looms began to fuel a new cotton manufacturing dynasty located around Manchester and Lancashire. Massive textile mills could now be built near population centers, freed from the need for rushing rivers to power the looms and spinners. In 1811, there were about 650 steam-powered cotton mills around Bolton, England. By 1841, there were more than a thousand mills spread across north central England, employing over 350,000 people. Over 70% of the mills operated around Lancashire, where one-third of the population was employed in the mills – men, women and children.

Life of the textile workers in the cities of England

The hundreds of thousands of textile workers now employed in the cities of England lived under terrible conditions. Arkwright had worked people (many of them children) terribly hard in his country mills, under very dangerous conditions, but he had provided healthy living space for his employees and was not overtly brutal in his management style. Life in and near the mills of Manchester and Lancashire, however, was a nightmare. As Steven Yafa (2005, p. 67) describes it: "The factory system . . . quickly gained notoriety as England's squalid national disgrace. Manchester and numerous other cities built around cotton textile industry became filthy, dehumanizing urban blemishes."

Displaced spinners and weavers flooded from the countryside to the cities in search of work. Hungry, desperate immigrants from Ireland and Italy also swelled the labor pools. The teeming mass of people far exceeded the number of available jobs, inducing ruthless factory owners to pay workers pitiful wages under unconscionable conditions. Orphaned children from parish wards and workhouses became particularly desirable employees, being docile and accepting pay far lower than adults. In 1830, over 35% of the 110,000 workers in Lancashire were children, and another 200,000 children toiled in Manchester mills.

The ones "lucky" enough to find jobs in the mills were typically subjected to fourteen-hour work days under cruel bosses. They started their day at 5:00 on an empty stomach, and worked for several hours before they were brought a cup of weak porridge and an oat cake, which they had to wolf down without stopping work. They had a short lunch break at noon and then toiled continuously until 8:00 or 9:00 that night without any additional nourishment. There was absolutely no rest for the weary, as the exhausted workers were given no additional breaks and had to remain standing at their workstations all day long. The floor bosses loudly cursed slow workers, sometimes beating them with sticks. The few children who survived this ordeal were often crippled for life from the constant bending in unnatural positions.

The atmospheric conditions they worked under were simply abominable. Windows were kept closed to keep humidity high so that the cotton yarn remained pliable with minimal breakage. The air was filled with cotton dust and a range of nauseating smells, ranging from wet cotton, human sweat and engine oils. Many workers developed brown lung disease (byssinosis), a scarring of the lungs that left them gasping for air. The loud, reverberating pounding of machines left many workers deaf from the relentless thunder. Surrounding the workers were countless whirling wheels and cogs ready to capture fingers and hands. Children sent to scavenge loose cotton floating about the plants were often crushed by operating machinery.

It seems hard to believe, but their pitiable work conditions were rivaled by their cramped, filthy living conditions. The workers lived in slums, up to ten people packed into rooms in damp, filthy houses. Sanitary conditions were abysmal, and the scores of people living in a hundred houses might share a single privy that was just an open pit dug in a corner of a yard. Windowless basements, where the poorest of the poor had to live, often were flooded during heavy rains with foul, fetid runoff. The billowing coal smoke from industry and home made the air acrid and the rain acid.

All known respiratory diseases were rampant. Thousands suffered from bronchitis and asthma, and many succumbed to influenza, tuberculosis and pneumonia. A cholera epidemic in Manchester in 1831 took hundreds of lives. Many of those too weak to work anymore simply lay waiting to die, hungry and homeless, in muddy doorsteps and alleyways. Most mill workers lived only a few miserable years after employment, with children rarely surviving their teens.

Cotton cultivation moves to the New World

By the 1800s, the United States became the major supplier of raw cotton to the burgeoning mills of England. Cotton cultivation had begun moving to the New World in the mid-sixteenth century, first following sugar cultivation to the Cape Verde Islands and then on to the West Indies. Planters in Virginia were growing cotton in the 1640s and in South Carolina in the 1660s (Riello 2010). South American plantation owners in Suriname began planting cotton in about 1730.

Cotton production in the West Indies was dwarfed by sugar production, but gained a toehold as a means of diversification. Cotton was bulky and expensive to ship, and sugar (and rum) production was much more profitable. On St. Thomas, 80% of the plantations grew some cotton, but bales of cotton were only about 2%–4% of the total plantation exports (Riello 2010).

The first cottons widely grown in the southern states of the United States were Sea Island types of *Gossypium barbadense* imported from the West Indies (Stephens 1976b). The Sea Island cottons may have been introduced into Barbados from Peru in the early 1600s. Speculation abounds, but it is possible that the Sea Island cottons hybridized with local wild upland types of *G. hirsutum* before they found their way to the United States. The original Sea Island cottons grown in the West Indies were short day types, while those grown in South Carolina and Georgia carried the day-neutral trait found in the upland types. Feral races of both types can still be found in the West Indies.

Cotton growing in the New World began to really pick up steam in the last quarter of the seventeenth century, when the British cotton mills began to roll out significant amounts of cotton yarn and textiles. The West Indies and Brazil were the first major suppliers of raw cotton to England, with Jamaica leading the way. The southeastern United States took command in 1800, producing about 50% of the cotton that fed the British mills. By 1820, this proportion had risen to about 80%, where it remained until the Civil War. The West Indies had abandoned cotton growing altogether by 1835 and Brazil was reduced to a minor player by that time.

As the British factories' insatiable need for raw cotton continued to grow, US cotton production keep pace by expanding from the original British colonies of South Carolina, Virginia and Georgia into the vast, rich Mississippi Delta (Johnson 2013). To support this expansion, the Native American occupants were ruthlessly eliminated and the huge, now surplus slave populations of the southeastern states were relocated. As Walter Johnson (2013, p. 5) describes it, "White privilege on an unprecedented scale was wrung from the lands of the Choctaw, the Creek, and the Chickasaw and from the bodies of the enslaved people brought to replace them." As many as a million enslaved people were forcefully redistributed across hundreds of miles in a vast domestic slave trade.

It was largely Andrew Jackson who directed the ethnic cleansing of the southeastern quarter of the United States. From 1817 to 1840 he supervised the depopulation of the region, first as a general in the US Army, then as military governor of Florida and finally as president of the United Sates. He started the process with an illegal war with the Spanish in Florida, then took on the Seminole Nation, and moved steadily west subjugating the Chickasaw, Choctaw and Cherokee. All tribes suffered the same fate – an offer of "equal land," which ultimately became a forced relocation to the Indian Territories (Oklahoma). Indian leaders had no other option after protracted running battles had left their tribal strength weakened to the point of certain defeat. Thousands died en route in "trails of tears."

As the Native Americans were driven out, the cotton plantations moved in. Cotton profits spurred westward expansion, as planters sought new opportunities and rich soils to replace their exhausted ones. The Mississippi Delta also had superior inland waterways compared to Georgia, Virginia and South Carolina. The path to southern aristocracy was to own both land and slaves, and cotton provided the profits to obtain more and more of both. By the end of the antebellum period, there were more millionaires per capita in the Mississippi Valley than anywhere else in the United States.

To provide the Cotton Kingdom's workforce, probably a million enslaved people were forcefully redistributed across hundreds of miles in a vast domestic slave trade (Johnson 2013). Slavery was given an unfortunate rebirth. The domestic slave trade began with individual speculators who would purchase slaves on credit in Virginia and Maryland, and then march them, chained wrist to wrist, first to Georgia and later to Louisiana, Mississippi and Alabama. Soon, huge slave trading firms emerged in the South with fine offices, show rooms and jails where up to a hundred slaves could be kept captive.

Major slave trade routes developed: coastal ones from Upper South ports to New Orleans and overland ones starting in New Jersey and fanning out from places like Richmond, Virginia, and Fayetteville, North Carolina, to Florence, Alabama, and Jackson, Mississippi. Agents were hired to scour the countryside for available human chattel. At least half of the slave purchases broke up families. Slaves were classified by their sex, age and condition; their hands were considered to be especially important. Long, nimble fingers were thought to be best for the picking of cotton.

By 1850, about half of the total US cotton crop was grown by slaves in just three states, Alabama, Mississippi, and Louisiana. In 1800, there were 100,000 slaves in what is now Mississippi and Louisiana; 250,000 in 1840; and 750,000 in 1860. Rich forests had been cleared to make way for cotton growing, and abandoned when the soil was exhausted. Most of the plantations were large-scale operations – an optimal size was considered to be 1,000 to 1,500 acres, with 75–100 slaves. Cotton growing was unique among the plantation crops, in that only the raw product was produced; there was no processing or refining. Most food and provisions were bought and shipped from the North, as cotton was too profitable to grow anything else.

Revolutions in cotton growing

With the expansion west of cotton growing came a great change in the types of cotton grown (Moore 1956). In the late 1700s, in the Carolinas and Georgia, the Sea Island cottons were cultivated. These were not particularly high yielding, but had exceptionally long staples that could be pulled away from the seeds with relative ease. Unfortunately, they were not well adapted to the climate of the Mississippi Delta, suffering from frosts and the prevalent saline soils. Much higher yielding in the Deep South were the Mexican upland varieties of *Gossypium hirsutum* that had been introduced in the 1700s.

The early varieties of upland cotton had two major problems – they were hard to pick and had a short staple that was extremely difficult to separate from the seeds. The major breakthrough came when Eli Whitney invented the cotton gin in 1793, which made the removal of seeds from the existing Mexican upland varieties quick and easy. Before the cotton gin, it took a slave a whole day to extract the seeds from a pound of cotton; with Whitney's gin the same person could do fifty pounds in a day!

Similar to most of the English textile mill inventors, Whitney made very little from his breakthrough invention. He and his partner, Phineas Miller, decided that they would make their profit by building cotton gins and then licensing them to planters for one-third of their crop each year. This proved to be a dismal failure, as a competing design soon emerged, and

Figure 3.2 Whitney's cotton gin

Source: US Library of Congress.

the growers found it easy to copy his machine without paying any royalties. Whitney was able to win some cash awards after more than a decade of claims, but the awards never matched the real value of his invention. All told he received $90,000.

The next major breakthrough for Mississippi Delta planters came in 1820, when the first really good upland variety, Petit Gulf, was developed in Rodney, Mississippi. Its genetics were based on upland cotton seeds smuggled inside Mexican dolls by Walter Burling of Mississippi in 1805. Petit Gulf had very high yields, a much longer staple than its predecessors and was prized for its "pickability." Before Petit Gulf, the Deep South relied on what was called "Georgia Green Seed cotton," introduced in 1734 from the West Indies and "Creole Black Seed cotton," reportedly introduced from Siam in the 1730s. Creole Black Seed had very small bolls that were very difficult to pick, and Georgia Green Seed had lint that hung tenaciously to the seed, making ginning extremely difficult.

Over the next several decades, a series of improved varieties appeared across the South that were selected by the growers themselves. Rather than plant all the seed from the previous harvest, they learned to select only the uniform, well filled and healthy seed, a process that purified their seed stocks. In the 1840s, Henry Vick became the most successful cotton seed producer in the Deep South, when he began to employ the selection methods used by northern corn breeders. Instead of just selecting superior looking seed from his bins, he went a step further and selected the best looking plants in the field. He would send a group of slaves out ahead of the picking gangs to gather seed cotton from only the most productive plants. Vick made a healthy profit from his seed, although he was widely ripped off by unscrupulous seed dealers including his own neighbors.

The treatment of slaves in the Cotton Kingdom

The life of southern slaves in the early 1800s was one of unceasing toil and relentless degradation (Northup 1853; Hebron 1988; Davis 2006; Johnson 2013). They had no freedoms at all, were regularly beaten bloody by the whip, given only enough food to keep them alive, and were forced to work long tortuous hours with little break. The only days they had off from toil were Sundays when they were not picking cotton and several days around Christmas when the most benevolent planters might give them a feast.

Many lived in packed log cabins with dirt floors, and their possessions were often limited to the rough cotton clothes on their back and the dirty blanket they slept upon. Their evening sustenance was often only a slice of bacon and a handful of corn meal moistened into a patty. They cooked in open fireplaces with chimneys made of sticks and mud. Many didn't have a pot or pan, so the food was cooked directly on the fire with the ashes being brushed off after preparation. They often slept on the bare ground or perhaps on a board, wrapping themselves in their rough blanket. Cold wind and rain often streamed between the logs of their cabin, making sleep difficult.

As David Brion Davis (2006, p. 196) put it, "plantations in the Deep South were essentially ruled by terror." It was not uncommon for the strong willed to be savagely beaten for sport. Those deemed the most incorrigible might be stripped and hung for hours by their hands with their toes barely touching the ground. Others might be forced to lie face down, with their hands and feet tied to posts, and subjected to hundreds of whip lashes. Hired patrollers roamed the countryside seizing slaves thought to be runaways, often siccing packs of dogs upon them. Women slaves were raped at will by lusty planters and overseers. The sad truth was that when babies resulted from these forced unions, the perpetrators had turned their sperm into profit.

The sad rhythm of slave life

In the early spring, the cotton fields were planted by gangs of slaves (Northup 1853; Hebron 1988; Davis 2006; Johnson 2013). In the lead were mule drivers who built ridged beds using plows. Behind them children followed who dropped seed from big baskets hung around their necks. Bringing up the rear were more mule drivers with harrows, throwing soil over the seed. Everyone was required to work at a feverish pace and given little break, even to eat their meager lunch.

From April to July, gangs of women and men were sent to the field to hoe weeds and thin cotton. The weeds between the rows were "scrapped" using mule driven plows, while the weeds within the rows were manually hoed. The cotton was also thinned, leaving only the strongest plants at distances of two to three feet apart. The slaves raced as fast as they could along the row, bent over, not standing until they finished a row.

The slaves were required to work in lines, with the fastest worker always at the front. Black "drivers" followed with whips cracking, trying to keep everyone in pace with the leader. As Solomon Northup describes in *12 Years a Slave*:

> If one of them passes him [the leader], he is whipped. If one falls behind or is a moment idle, he is whipped. In fact, the lash is flying from morning until night, the whole day long. . . . It is the literal unvarnished truth

> that the crack of the lash, and the shrieking of slaves, can be heard from dawn to bedtime.
>
> (p. 126)

In the latter part of August, the harvest began and continued into December. The slaves dragged long sacks strapped to their necks with an opening about breast high. Larger baskets were placed at the end of the rows to empty the filled sacks. Every slave had to meet a daily picking quota. Two hundred pounds was considered an acceptable amount for an experienced picker, but those who could pick more were held to it. As Northrup recalled,

> When a new hand, one unaccustomed to the business, is sent for the first time into the field, he is whipped smartly, and made for that day to pick as fast as he can possibly. At night it is weighed, so that his capacity in cotton picking is known. After that this becomes his benchmark – he must bring in the same amount every night. If it falls short, it is considered evidence that he has been a laggard, and a greater or less number of lashes is the penalty.
>
> (pp. 115–116)

The picking each day did not stop until the overseer called a halt when it was too dark to see. On bright, moonlit nights they might have to continue until midnight. The slaves were then required to carry their cotton to the gin house to be weighed. This was a time of great dread, because if they had picked too little they would be beaten, and if they picked too much they would have a new standard to bear. When the whippings were over, they still had to complete other chores, such as cutting wood and feeding the mules, before they could eat and get what rest they could. A horn sounded an hour before dawn to wake everyone and woe to the slave that was not in the field by dawn. The lash awaited them.

Not all plantation slaves were treated so harshly as described. Some rare planters shunned the whip and instead provided cash rewards for high productivity. Task systems evolved in South Carolina where slaves were given specific goals and when completed were allowed to relax or garden. As cotton fields became more mechanized in the 1840s, many slaves became skilled operators of horse-drawn seed planters, cultivators and scrapers. More land was devoted to growing food crops, and herds of beef and dairy cattle were maintained, affording both slave and owner with a better diet. Husband and wife were encouraged to live together and tend their own gardens and livestock. Chairs and beds began to appear in cabins. However, even the most fortunate cotton laborers were still held in degrading bondage. They did not have the "certain unalienable Rights" mentioned in our Declaration of Independence, had no choice but to do the overseer's bidding and were considered as property, not human beings.

A US cotton textile industry emerges

As cotton plantations began to greatly expand across the southern US states in the decades surrounding 1800, a textile manufacturing industry also emerged in the northeastern United States (Batchelder 1863; Mitchell 2008; Rosenberg 2010). The first mill was started by Samuel Slater in 1790 at Pawtucket, Rhode Island. A former superintendent of a cotton mill in England, Slater had sneaked into the United States disguised as a farmer. English customs officials were not allowing anyone who had worked at a mill to leave the country, in an attempt to keep their monopoly on cotton textile production.

Relying only on his memory of Arkwright's water frame, Slater managed to build a yarn factory with the help of local blacksmiths and mechanics. He got financial backing from the merchant Moses Brown, a Quaker and strong abolitionist, whose family money had come partly from the slave trade. The cotton used to feed the mill came first from Brown's West India farms and then later from Carolina. Slater's mill spun only long staple Sea Island cotton, as his machinery was too crude to spin the shorter staple upland cotton.

Textile mills sprang up all across New England over the next couple of decades. By 1800, there were seventeen cotton mills and 14,000 spindles in Rhode Island and Massachusetts, and by 1812, there were thirty mills and almost 50,000 spindles. Across the whole of New England, from Rhode Island to Maine, there were about 125 mills by 1810 and over 200,000 spindles. All these mills were producing only yarn; weaving was still left to the home industries.

The first US mills that both spun yarn and wove cloth were established by Francis Cabot Lowell in 1814. Lowell, a Harvard graduate and member of an aristocratic Boston family, had gone on a grand tour of Europe for his health and while there got the idea to make his fortune building textile mills in America. Through family connections, he was introduced to a number of English textile mill owners, and they incredibly gave him and his Boston friend Nathan Appleton complete access to their mills and engineers. Lowell first tried to license textile machinery and received a resounding "no," forcing him to commit the whole textile process to his photographic memory.

On his return in 1812, he immediately began to search for financial backers. The timing was perfect, as the War of 1812 had severed Anglo-American trade and stimulated local investment in manufacturing. He got some support from Appleton, but the bulk of his startup monies came from his brother-in-law Patrick Tracy Jackson and his two brothers. The group put up a total of $400,000 to establish what Lowell named the Boston Manufacturing Company.

With the help of an ingenious mechanic Paul Moody, Lowell was able to draw on his memory to build a textile mill, which opened two years later at

Waltham, Massachusetts (about ten miles from Boston). Their fabric proved to be costlier to produce than that from England, but the fledgling company got a great boost when the federal government passed the first tariff in 1816 to protect American industry.

Lowell was able to enjoy its success for only a few years, dying at the age of forty-two in 1817. The management fell to Jackson, who moved the whole operation to the Merrimack River near Concord in 1821. Under his direction, what was now called the Merrimack Manufacturing Company became an empire. The company bought control of the Pawtucket Canal and the surrounding farmlands and established the city of Lowell to house its workers.

By 1823, the mills of the Merrimack Manufacturing Company spread over a mile along the Merrimack River. The city of Lowell had 16,000 occupants, of which 6,000 worked in the mills. The workers lived in company dormitories, purchased their goods in company stores and worshipped in company churches. Lowell came to be called the "City of Spindles" or "the Manchester of America."

When the Lowell mills went through a rough financial period in 1830, the wealthy brothers Amos and Abbott Lawrence became financial backers. They owned the greatest wholesale mercantile house in the United States. Under the tutelage of the Lawrence Brothers, along with Jackson and Appleton, the empire was further expanded to thirty-one mills in Massachusetts, New Hampshire and southern Maine by 1845. These "Boston Associates" amassed huge amounts of wealth, taking over railroads and establishing banks. Their mills came to produce one-fifth of all the cotton textiles in the United States, with annual profits averaging over 20%.

Life of the textile workers in New England

The majority of the mill workers in New England were women, mostly from farms. It was the first time that unmarried women could earn a living outside the home (Yafa 2005; Hall *et al.* 1987; Anonymous 2014a). They were not treated as cruelly as the textile workers in Britain, but they suffered the same mind-numbing tedium and unhealthy work conditions. They lived together in cramped company boarding houses adjacent to the mills. The women toiled long hours with few breaks in steamy, poorly ventilated rooms with the constant booming of machinery. They had a regimented life dictated by clanging bells. They were woken at dawn by a bell warning them to be at work by 5:00 a.m. and they stayed at their work posts until 7:00 p.m. Their only rest came at 7:00 a.m. and noon, when a bell signaled breaks to eat.

In spite of their brutal work schedule, many of the mill girls worked hard to better themselves. They joined self-improvement groups, attended Lowell

Hall lectures of dignitaries such as Ralph Waldo Emerson, and availed themselves of local libraries. An internationally acclaimed magazine called the *Lowell Offering* was established in 1840 that contained verse, essays and stories written by the mill girls.

The women workers did not suffer their brutal work conditions and pitiful pay without repeated protests. In one of their largest outcries after a reduction in their wages, about 1,500 women walked off the job in 1836 and gathered daily at a neighborhood park. There they heard rousing speeches and sang a sad, protest song called "I Cannot Be a Slave" (http://library.uml.edu/clh/All/so04.htm).

> Morning bells I hate to hear
> Ringing dolefully, load and clear
> Now isn't it a pity
> Such a pretty girl as I
> Should be sent to the factory
> To pine away and die
> But I cannot be a slave; I will not be a slave
> For I'm so in love with liberty
> I cannot be a slave!

The mill owners finally did rescind the salary reduction after the women had been on strike for several weeks, but it proved to be a rare victory in sea of losses.

A former mill worker named Sarah Bagley organized a labor reform society in 1844 with its primary goals to reduce the work day to ten hours and improve health conditions. She produced broadsides that railed against management and started the pro-labor magazine *The Voice of Industry*. She led a petition drive demanding that the Massachusetts legislature establish a ten-hour workday. An investigative committee was set up that interviewed Sarah and many other factory women, along with doctors testifying about rampant disease, but it all came to naught. In their report to the whole legislative body, the committee carefully chronicled all the problems at the mills but offered no recommendations for change.

In the last two decades preceding the Civil War, fewer and fewer New Englanders were willing to suffer the poor pay and indignities of the mill and they were replaced by immigrants, mostly Irish. These people had few options for work other than the mills and were desperate for any employment. Sarah Bagley and the mill girls had tried valiantly to reduce hours and make the mills a healthier place, but decades of protest had done little to improve conditions. In fact, pay was actually lower than two decades before and work hours had been reduced by only a half hour. Real change in the mill worker's conditions would have to wait until the next century.

The emergence of sweatshops

To make cloth into garments, small tenement "factories" sprang up throughout London's West and East Ends in the 1800s (Kingsley 1850; Riis 1890; Anonymous 2011). Charles Kingsley in his pamphlet "Cheap Clothes and Nasty" compared the clothing workers to slaves. They were brought from Ireland for a fee and became trapped in the life, never able to earn enough to go back. The middlemen who bought and sold the goods were called "sweaters," squeezing every drop of energy out of the miserable wretches at the sewing machines and ironing boards.

The garment workers of London lived and toiled in cramped, stifling rooms, often with ceilings too low to stand upright. They made a pittance in wages, out of which the owners extracted fees for their lodging and meals. It was not uncommon for the sweater, his wife and children and six workers to live and work in two rooms, the largest being eight feet by ten feet. There was no ventilation in the rooms and they continually breathed in foul, almost suffocating air. Many were consumptive.

Conditions were no better for garment workers across the ocean in New York, which became the center of the garment industry in the late 1800s. Following the wretched pattern established previously in England, the workers were mostly immigrants stuck in cramped, filthy tenement factories. They came to be known as the "sweaters of Jewtown" in Jacob Riis's 1890 exposé entitled "How the Other Half Lives." The sweaters battled each other in a cutthroat bidding war. As Jacob Riis described:

> His price is not what he can get, but the lowest he can live for and underbid his neighbor. Just what that means we shall see. The manufacturer knows it, and is not slow to take advantage of his knowledge. He makes him hungry for work by keeping it from him as long as possible; then drives the closest bargain he can with the sweater.
>
> (p. 38)

The sweater's livelihood depended on the "helpless ignorance" of a constant stream of desperate Jewish immigrants from Poland and Russia. As Riis put it:

> But as long as the ignorant crowds continue to come and to herd in these tenements, his grip can never be shaken off. And the supply across the seas is apparently inexhaustible. Every fresh persecution of the Russian or Polish Jew on his native soil starts greater hordes hitherward to confound economic problems, and recruit the sweater's phalanx. The curse of bigotry and ignorance reaches halfway across the world, to sow its bitter seed in fertile soil in the East Side tenements.
>
> (1890, p. 38)

Figure 3.3 A group of sweatshop workers at 30 Suffolk Street in New York, 1908
Photographer: Lewis Wickes Hine. Source: Library of Congress.

As in London, the work was done in small apartments lining dark, dank stairways in tenements. The garment workers started their toils in the early morning and worked long hours into the night, at least six days a week. Meals were consumed while working. Children wandered among the stacks of finished and unfinished garments, their bodies stained the color of the garments. When enough clothes had been produced to make a meager profit for the day, the workers and their families collapsed, often three abreast, into beds in the same room as their machines.

The crowded, abysmal conditions found in the tenement "factories" made them ripe for disaster. The whole nation was scandalized in 1911 when a great fire broke out in the Triangle Shirtwaist Factory in New York. The upper three floors of a ten-story building caught fire and 145 people died horribly in the blaze. The terrified workers had little chance to flee the flames as the single fire escape collapsed and the exit doors had been locked. Bodies rained down on the street below as burning women hurled themselves through windows, having no other means of escape. Piles of bloody, broken bodies littered the street.

Remarkably, the owners of the factory tenement got off with little punishment. They were acquitted of any wrong doing in a criminal trial, and in twenty-three civil law suits they paid an average of only $75 per life. The state legislature did, however, react to the public outrage associated with the disaster. They quickly established a Factory Investigation Commission, which identified 200 other sweatshops in New York where a similar tragedy could occur. The government passed thirty-six labor reform acts between 1911 and 1914 covering a wide spectrum of safety measures including ceiling sprinklers, improved eating and toilet facilities and shorter work hours. The tragedy also brought attention and membership to the International Ladies' Garment Workers' Union, which fought long and hard for improved sweatshop conditions.

The stranglehold of cotton on the US economy

By the early 1800s, cotton growing was king in the southern United States. A number of crops including indigo, tobacco and rice were still grown, but cotton predominated and had a stranglehold on the southern economy. Cotton essentially drove the economy of the South, and greatly influenced its social structure (Yafa 2005; Johnson 2013). In the years preceding the Civil War, cotton production and transportation employed the majority of the free workers and slaves of the South. Southern life was all about agriculture and had become centered on cotton.

Southern cotton also played an important role in the industrial growth of the North. More than half of the total value of all US exports came from cotton from 1815 to 1860, and cotton earnings paid for the majority of all imports by 1860. Northern merchants first made huge sums shipping cotton to European textile mills, and by the early 1800s began funneling those profits into their own textile mills using southern cotton. The textile industry in the North was producing $100 million worth of cotton cloth annually in the mid-1800s, and selling the South more than $150 million of manufactured goods. The cheap cotton clothing worn by the slaves of the South came from the northern textile mills. Cotton even played an important role in the western and midwestern parts of the United States, which provided southern cotton growers with the majority of their food each year.

Financing the cotton plantations of the South greatly entangled the planters with northern financiers. Every year, hundreds of Yankee "factors" spread across the South offering loans of 7–12% interest on the next year's crop. The planters were dependent on those loans to grow their crop, feed and clothe their slaves, and fill their houses with fine furnishings. Bales of cotton were payment, fluctuating wildly in price each year depending on world markets. Often a harvest that promised great wealth when it was brought from the field was dashed in value by the time the bales arrived in Liverpool. A sudden drop in cotton demand by the British mills in 1837 brought on a panic that ruined many southern planters and collapsed the whole financial system of the United States.

There came to be a national cotton aristocracy, composed of southern planters and the northern mill owners, merchants and financiers. They considered each other as valued peers and became integrated socially. Many sons of southern planters studied at northern universities such as Harvard,

Yale and Princeton. They all did their best to ignore the sectional differences to protect their lifeblood, King Cotton. The powerful Boston Associates took a hands-off approach on the question of slavery, preferring to maintain the status quo rather than rock the boat by acknowledging the howls of the abolitionists.

Foul winds rising

The economies of the North and South were both dependent on cotton, but it took many contentious compromises on slavery to hold the Union together (Yafa 2005; Bordewich 2012). The process began at the Philadelphia Convention in 1783, when James Wilson proposed that three-fifths of slaves should be counted as "persons" for determining taxes and representation in the House of Representatives and Electoral College. Delegates from states with few slaves had wanted the slaves to be included in taxation but not in determining representation, while delegates from the slave states wanted the opposite. The ratio of three-fifths had no logical basis except as a compromise between proposals of one-half and three-quarters.

For over thirty years, the topic of slavery remained dormant in the House and Senate, even in the face of growing anti-slavery sentiment in the North. No one had a clear idea how the slaves could be freed without catastrophic effects on the whole US economy, and it was deemed best to simply leave the topic in the shadows. The issue finally came to a head in 1820 when Missouri, with its sizable slave population, applied for statehood. At this time, the Senate was balanced with equal numbers of Representatives from free and slave states (11), but only 40% of the members of the House of Representatives were from slaveholding states. It was clear to all the legislators that the inclusion of Missouri in the Union as a slave state would greatly strengthen the Southern hand.

A ferocious and prolonged battle ensued. The first salvo came from Representative James Tallmadge of New York who wanted to ban the further importation of slaves into Missouri and free children born to slaves at age 25. This proposal barely got through the House but was soundly defeated in the Senate. Senator Jesse B. Thomas then proposed that Missouri be allowed to enter as a slave state, but that slavery be excluded in the rest of the Louisiana Territory north of latitude 36′30″. To keep the balance of power where it was, he proposed that slave-free Maine also be granted statehood. After more contentious debate, patiently and tirelessly moderated by Henry Clay, the measure passed with a small majority of Southerners and a few Northerners.

The acrimonious fight did not end here, as participants in the Missouri Constitutional Convention legalized slavery as expected, but went a step forward and also banned free Negros and mulattos from entering the state. This clause was clearly unconstitutional, and while it got through the US Senate, it was rejected handily in the House of Representatives. To garner the support of both legislative bodies, Henry Clay proposed that Missouri be required to remove the offensive clause, but suggested as a compromise the requirement that only American citizens would be admitted into the state. This proved acceptable to the Senate, and the Missouri legislature was directed to drop its illegal ban, which it did a few months later.

This contentious debate between slave and non-slave states sounded an alarm that the Union could be blown asunder over the issue of slavery, regardless of cotton's central role in the southern and northern economies. The aged Thomas Jefferson wrote, "This momentous question like a fire bell in the night, awakened me and filled me with terror. I considered it at once as the knell of the Union."

South Carolina was the first state to threaten secession in 1832, when it asserted that any state could "nullify" a federal law that they did not like, in this case the federal tariff. President Andrew Jackson, even though he was himself a slaveholder, angrily threatened to lead 50,000 federal troops into South Carolina to enforce the federal laws with the support of Congress. Steady old Henry Clay again defused the situation by proposing a compromise tariff that placated the South Carolinians.

In an attempt to prevent future flare-ups in the Congress, the House passed a "gag rule" in 1836 that prohibited the discussion of any outside petitions. Petitions of all kinds were now flooding into the House and Senate, with many in opposition to slavery. One of the most volatile came from the American Anti-Slavery Association, which wanted to abolish slavery in the District of Columbia. In national campaigns, they had amassed 34,000 signatures in 1836 and 87,000 in 1838. The gag rule kept discussion of slavery at bay for nearly a decade until it was finally repealed in 1844 through the tireless efforts of John Quincy Adams, once a president and now a representative.

The Union teeters at the brink

By the mid-1800s, it had become crystal clear to southern leaders that their strength in the federal government was dependent on the expansion of slavery into the territories. Abolitionists such as William Garrison, Frederick Douglass and Harriet Beecher Stowe were turning multitudes of Northerners against slavery and it was becoming increasingly difficult to find northern support in the Senate and House of Representatives. The ultimate survival of the "peculiar" institution of slavery would be dependent on maintaining the balance of power between the slave and free states (Bordewich 2012).

The issue of expanding slave territory was brought to the forefront in 1849, when an obscure Pennsylvania Democrat named David Wilmot proposed that slavery be prohibited in the vast new territories obtained in the Mexican War. The "Wilmot Proviso" came at the same time that California requested admittance as a free state. Defiant John C. Calhoun, a former vice president and secretary of state, led a hostile response that threatened secession of the whole South. Both Northern and Southern legislators feared that the nation was about to collapse.

The venerable Henry Clay, now gravely ill with tuberculosis, decided to try his magic one more time. In the Compromise of 1850, he proposed a bundle of five bills: (1) admission of California as a free state; (2) the organization of Utah and New Mexico into states that could decide for themselves on the issue of slavery; (3) the payment to Texas of $10 million for the land it claimed in New Mexico; (4) the exclusion of the slave trade in the District of Columbia; and (5) the Fugitive Slave Act, which would require federal marshals to return runaway slaves.

No representative favored more than a few of these proposals, but Clay felt that they had to be passed together as a grand compromise. There was no longer any single proposal that could save the Union. Acrimonious debate raged back and forth for nine months. President Zachary Taylor vehemently opposed the plan, but his opposition dissolved when he died. Calhoun became increasingly weak and also died. Passage of the last of the bills finally came about when the great Massachusetts orator, Daniel Webster, argued in support of the Fugitive Slave Act. This bill was bitterly opposed by the abolitionists in the North, and Webster's support of it ended his dreams of being president.

The Compromise of 1850 kept the Union together for a few more years, but the ugly battle renewed over the statehood of Kansas in 1854. Senator Stephen A. Douglas proposed that the Missouri Compromise be rejected, and that the people in the territory be allowed to decide whether they were slave or free. He called this doctrine "popular sovereignty." Old wounds were opened and violent discussion ensued. Free and slave supporters flooded into the territory and people were murdered over the issue. Douglas put up a stout defense and the proposal eventually passed: 37 to 14 in the Senate and 113 to 110 in the House.

The Union still lived, but not for long. In May 1856, the anti-slavery Senator Charles Sumner of Massachusetts gave a particularly venomous attack on the South in the US Senate, known as "The Crime against Kansas." In it, he lambasted the South on the issue of slavery and singled out South Carolina's Senator Andrew Pickens Butler, suggesting that his mistress was the "harlot slavery" and he "touches nothing that he does not disfigure with error." The next day Butler's cousin, Representative Preston Brooks, sneaked up behind Sumner and proceeded to savagely beat him with the gold head of his cane until Sumner fell to the floor unconscious and bleeding profusely about his head. It took him years to recover. This event so shocked the Boston Associates that they realized they could no longer vacillate on slavery. The South had lost its most powerful supporter in the North. The last of the cement holding the North and South together had evaporated.

The Republican Party emerged in 1856 as abolitionist sentiments fractured the old Whig party into northern and southern pieces. The major part of the Republican Party's platform was to keep slavery out of the new states and territories, a view now supported by the Boston Associates. Lincoln became its presidential candidate in 1860 and won the Electoral College with no southern support. His election made the slaveholding states face the bitter reality that they could no longer sway national power, and it was only a matter of time until slavery would be abolished or at least severely restricted. National power could now be bestowed on a party without their support. Seven states seceded before Lincoln was inaugurated and another four soon followed.

Civil War and cotton diplomacy

The Civil War began with the South Carolinians firing on Fort Sumter in Charleston Harbor on April 12, 1861, and ended almost five years later at village of Appomattox Court House on April 9, 1865. By the time the war came to an end, over 600,000 Union and Confederate soldiers were killed by bullets and disease. Large swaths of the South lay in ruins and their slaves had been declared free. "Father Abraham" had given the slaves in the war zones freedom through his Emancipation Proclamation in 1864, and the rest of the slaves in the United States were freed by the Thirteenth Amendment in 1865.

As the Civil War dragged on, the planters in the South found it increasingly difficult to produce cotton (Yafa 2005; Johnson 2013). Most were fighting and could not supervise its planting and harvesting, and without northern trade, much more land had to be devoted to the production of foodstuffs. The South managed to produce 4.5 million bales in 1861, but this plummeted to 1.5 million in 1862, 500,000 in 1863 and only 300,000 in 1864. The value of cotton skyrocketed in Great Britain and France as they struggled to find alternate sources. Blockade running became worth the risk, but not nearly enough cotton could be smuggled out of the Confederacy to feed the hungry textile mills of Europe and the Northeastern United States.

The Confederacy had been convinced at the onset of the war that France and England would support them to save their own cotton-based economies. They could use financial help, but even more importantly, they needed naval support to break the Union blockade around their coastal cities and ports. The governments of Britain and France came very close to supporting the Confederacy when it looked like they might prevail, but when the tide turned at Antietam, they finally rejected the idea. The countries' overall revulsion toward slavery was too great to gamble on saving the Confederacy, even if it meant the textile mills would starve.

Effect of the Civil War on the textile industry

The Civil War had a devastating effect on the English textile industry (Fite 1910; Yafa 2005). Surpluses of raw cotton allowed the mills to run at full steam at the beginning of the war, but as the conflict dragged on mill after mill went bankrupt. By 1862, in what has been called the Lancashire Cotton Famine, almost half the spindles in Lancaster were idle and nearly a half a million people were being fed and clothed by charity. Significant amounts of dirty, short staple cotton began to arrive from India by the middle of the war, but its arrival was not in time to save many mills and hundreds of thousands of jobs. At the end of the war, almost one-third of the mills in England had closed.

The Civil War had a severe impact on the US mills, but the situation was not nearly as bad as in England (Yafa 2005). While many mills had to shut down for lack of cotton, some owners had bought huge quantities of cotton at a good price in 1860 which allowed them to stay open during the whole war. Other cotton mills were converted to woolen factories to supply the burgeoning army with uniforms, blankets, underwear and overcoats. Still others limped along at one-quarter to one-half capacities and survived on the high prices paid for cotton goods. Many of the men who did lose their jobs joined the military, and many of the unemployed women and children returned to their country homes.

Postwar shifts

When the war ended, the grand cotton estates of the South were in shambles. The once proud aristocrats still had their vast plantations, but they had no slave labor to work them and no money to hire employees (Yafa 2005). The only way they could see to make a living was to break up their war-ravaged plantations into parcels and rent them out to tenants or sharecroppers. Tenant farmers paid the landowner rent for farmland and a house. They owned the crops and after harvest, they paid their rent out of their profits. Sharecroppers owned nothing and borrowed for everything including their supplies, mules, plows and seed. They grew what the landholder told them to grow and after harvest the landowner sold the crop and applied the income to the sharecropper's account.

Poor whites who had vowed never to grow cotton were sucked up into the system, along with the former slaves. About 40% of the cotton was picked by whites, many missing limbs from the war. One-fifth of the Mississippi tax revenues in 1866 went for artificial limbs. In the late 1800s, poor crop yields and unstable prices drove more and more once independent farmers into tenancy. By 1890, one-third of all white farmers and three-quarters of all black farmers were either tenants or sharecroppers.

The sharecroppers lived in a feudal system of peonage, with little chance of ever getting out of debt. The profit from their crop went mostly to the landowners, and the rest often wound up in the pockets of the merchants who provided everything they needed on credit from farm implements to food and clothing. Almost all the banks had failed in the South, making these country merchants the only source of money. They kept careful ledgers of everything on credit, charging 30% to 70% more than the items were worth and loaning money at exorbitant interests of 20% to 100%. It was a system of legal larceny resulting in permanent debt for almost all sharecroppers.

By 1880, the South was producing as much cotton as before, but at far less profit. Cotton prices had plummeted to as low as fourteen cents a pound. The country was wracked in a decade-long depression, brought on by the "Panic of 1873," when the whole banking system of the United States collapsed through too rapid railroad expansion, lingering war debts and the failure of bonds. Many large landowners in the South had to forfeit their land as they could no longer pay their taxes.

Southern mill boom

One of the few economic bright spots for the South after the Civil War was the emergence of a textile industry (Copeland 1917; Yafa 2005). New steam-driven technologies had allowed the mills to move away from waterways, and improved spinning and weaving technologies had reduced the skill level needed at the mills. Mills began to crop up along the whole southern arc of the Piedmont encompassing parts of central North and South Carolina, Georgia and Alabama. By the turn of the century, the cotton textile industry of the South was close to matching the output of New England.

Similar to the early work force of the northern mills, the workers in the South came from surrounding farms (Hall *et al.* 1987; Hindman 2002; Fishman *et al.* 2014; Leloudis and Walbert 2014). Whole families of yeoman farmers came to work in the mills. They were originally subsistence farmers

Figure 3.4 A former farm family working in a Georgia cotton mill in 1909

Photographer: Lewis Wickes Hines. Source: US Library of Congress.

who had bartered for their non-food needs, but now were forced to grow cash crops like cotton and tobacco to pay new taxes levied to build roads and railways. The farmers traded their mules and plows for the promise of steady income in the mills.

The income was steady but poor, and most families had to pool their wages to make ends meet. The salaries were substantially less than even the pitiful wages of the New England mills. In the 1890s, an adult might make forty to fifty cents a day in North Carolina, while a child made ten to twelve cents. They worked sixty-hour weeks in the same hot, humid, noisy and dangerous conditions as in the northern mills, although the pace was a bit slower with some conversation and breaks. African Americans were excluded from all but the most menial, heavy jobs.

By 1900, most families lived in mill villages owned by the company. The villages consisted of a supervisor's home, houses for workers, a company store, schools and churches. People drew water from a common source and the sanitary system consisted of privies. Families generally went to work together, with one-quarter of the mill workers being children under the age of 16. Nursing infants were brought to work by their mothers, who fed them during factory breaks. Most children became full-time workers by the age of 12.

The struggle for mill children

Beginning in the late 1800s, many campaigns were undertaken to abolish child labor in the mills, but it took many decades of struggle to eliminate it (Hindman 2002; Fishman *et al.* 2014). Children's numbers gradually fell in the northeastern mills from 14.1% to 6.0% from 1880 to 1905, but in the Southern factories the proportion held steady at around 25%. The state legislatures were reluctant to take on the powerful mill owners and few laws were enacted to protect children, except setting age limits of thirteen or fourteen. Even these pitiful laws were rarely enforced, with desperate mothers lying about the age of their children to avoid starvation.

The great labor organizer, "Mother" Jones, was the first national leader to raise awareness about the plight of the mill children. In 1903, she led a march of crippled children from the mills of Kensington, Pennsylvania, to the home of Teddy Roosevelt in Oyster Bay. Many of the children were missing fingers and limbs, and were stooped and skinny from their oppressive work conditions. Roosevelt refused to meet with the group, but the march put a national spotlight on the plight of child laborers.

A mass meeting of concerned citizens was held at Carnegie Hall in 1904 to discuss the child labor problem. Out of these discussions came the National Child Labor Committee (NCLC), which was charged with raising national awareness of the conditions of child workers and lobbying state and federal legislatures on their behalf. The NCLC was composed of a glittering array of prominent clergyman, philanthropists, politicians and educators including a former president, Grover Cleveland, and the president of Harvard University, Charles W. Eliot. The NCLC was chartered by Congress in 1907.

One of the first things the NCLC did was to hire anthropologist Lewis Wickes Hine to pictorially chronicle the trials and tribulations of the child textile workers. For the next ten years he published thousands of images that pulled on the heartstrings of the American public. He was not afraid of bluffing his way into factories, facing great danger when his true intent was discovered. His riveting photos were a stark reminder that long hours in the factory robbed children of any future, irreversibly weakening their bodies and cheating them out of an education.

Even though much of the public was horrified by Hine's pictures, state and federal legislators remained slow to enact child labor laws. The pressure of the monster textile industry was simply too great. The NCLC did successfully lobby Congress to charter a Children's Bureau in the US Departments of Commerce and Labor in 1912, but it gained little traction in eliminating child labor and making children's education compulsory. Federal laws were passed in 1916 that limited the age and hours of child workers, only to be deemed unconstitutional by the Supreme Court. An attempt to skirt this roadblock by getting a constitutional amendment that would ban child labor failed to gain passage by enough states.

Finally, in 1938, after thirty-one years of struggle, the NCLC got its crowning victory when the US Congress passed the Fair Labor Standards Act that was accepted by the Supreme Court. This Act prohibits any interstate movement of goods produced by oppressive child labor, defined as work done by children under the age of sixteen or particularly hazardous work done by children from ages sixteen to eighteen. This Act remains today as the primary legal tool protecting the rights of children.

US mill worker strikes with little gain

Dissatisfaction among mill workers about their pay and working conditions finally boiled over in the early 1900s (Copeland 1917; Hall *et al.* 1987; Hindman 2002; Leloudis and Walbert 2014). A major strike of 25,000 workers took place in Lawrence, Massachusetts, in 1912 and lasted more than two months. A law had been passed by the state legislature that reduced the maximum hours that could be worked from fifty-six to fifty-four, so the owners had responded by reducing pay equivalently. Polish female weavers and many other immigrant groups rose up in objection, surprising the owners and union organizers alike. In the past, immigrants had been the quietest of workers and been used to break strikes. The strike led to congressional hearings of the abominable conditions in the mills, which spooked the owners into settling the strike and giving the workers a 20% raise.

During World War I mill production soared to supply the military, but after the war it dropped dramatically. Foreign competition also began to grow, and skirts got shorter requiring less material. What has been called the "Great Textile Depression" began in 1923. To maintain profits, owners reduced wages and began running the mills around the clock. They installed faster equipment, demanded their workers work harder and faster, and eliminated many jobs. Management started to carefully track employee productivity and tied wages to higher and higher quotas. The mill hands called these efforts to increase efficiency the "stretch out."

The most skilled made a little extra money because of the stretch out, but most made less and worked harder and harder. The lowest paid female spinners were hit particularly hard, being replaced by more efficient machines. Many older workers without pensions were fired because they could no longer keep up with the machines. The mill villages became squalid and cramped, the façade of company benevolence being long shattered. The workers no longer had even a semblance of dignity and autonomy. The atmosphere was ripe for revolt.

Worker frustration grew throughout the 1920s and erupted in 1929. The crash of the stock market had led to even deeper wage cuts and more layoffs at the mills. The first major strike broke out in Elizabethton, Tennessee, on March 12, which was quickly followed by strikes in Marion and Gastonia, North Carolina, and across the whole Piedmont. Violence erupted between the mill workers and city police in Gastonia, with the police chief being murdered. A local heroine and strike leader, Ella May Wiggins, was ambushed and killed by a mob as she tried to attend a union meeting. She had infuriated the locals by working to bring African Americans into the union. Five men were eventually indicted for her killing but were acquitted after only a half hour of jury deliberation.

The strike collapsed after the workers had suffered over three months of crushing debt, intimidation and near starvation. They settled for modest gains in wages, some improvements in mill village conditions and a slight reduction in work hours. A further promise that union workers would not be discriminated against was quickly forgotten – pitiful rewards for so much pain.

In a weak attempt to regulate the cotton textile industry, the National Industrial Recovery Act was signed into law by Franklin D. Roosevelt in 1933. It formed a regulatory body called the Cotton Textile Board that was supposed to limit price competition among the companies, reduce overproduction, and give mill hands a fair minimum wage and a shorter work week. In reality the Board did little to alleviate these conditions, as it had no enforcement tools and was populated with mill owners who had little interest in altering the existing practices. The minimum wage became the maximum wage and thousands more workers were laid off. When the Board shortened the work week to forty hours, the owners responded by requiring the workers to produce as much as they had in fifty to sixty hours.

Frustration among the mill workers again boiled over in 1934. By now, tens of thousands had unionized and joined the United Textile Workers of America (UTWA). The strikes began in Alabama with 20,000 walking off the job. The strikers demanded recognition of their union, a minimum wage of $12, a work week of thirty hours and reinstatement of workers fired for union activities. The UTWA organized a national convention and the delegates voted to go on a national strike.

The General Textile strike of 1934 (on Labor Day) became the biggest strike ever in the United States, with 500,000 total participants. Georgia, with its 60,000 textile workers, became the epicenter. The strikers thought they had President Franklin Roosevelt on their side, but he was slow to support them, not wanting to offend southern leaders and lose their support of New Deal legislation. Southern governors declared martial law and called out the National Guard. After being re-elected, Governor Tallmadge of Georgia reneged on campaign promises and instructed the Guard to arrest all strikers and everyone who aided them. The textile workers were beaten, bayoneted and forced into detention camps. Five workers were beaten to death. The union strike leaders were arrested and kept isolated from each other.

After only three weeks, the battered and hungry workers began to drift back to work, and some mills were able to operate with skeleton crews. The union was overwhelmed and simply did not have the resources to support such a massive walkout. Roosevelt set up a mediation board that produced a tepid ruling that sympathized with the worker's plight, but made no real recommendations except to do more study. The overstretched union was forced to call this a victory, but in reality, it had achieved no real economic gains. Many union men and most strikers lost their jobs, in spite of a strike termination agreement that required that they be rehired.

Never again would there be a strike of this magnitude in the textile industry. Union activities also would decline in the southern mills, bucking a national trend. The Congress of Industrial Organizations (CIO) pushed hard to organize the southern textile industry in the 1940s, only to fail dismally after facing violent opposition from not only business leaders but many of the workers themselves. The TWUA also tried vigorously to grow its membership and failed. When strikes did break out in the South, they rarely lasted very long, as managers were able to find hundreds of more than willing strikebreakers. The plight of the mill workers remained firmly in the hands of the factory owners and managers.

Gandhi and the textile workers of Ahmedabad

In the early 1900s, the textile workers in Ahmedabad, India, also struggled mightily with the mill owners for a decent living (Sujata 1984; Choi 2012). For centuries there had been few major confrontations between the workers and owners, even though plant conditions were abysmal. The workers were finally riveted into action in 1918 when the owners took away salary increases that they had been given the previous year. The owners had raised salaries to keep their workers from fleeing the plague that had reached Ahmedabad in 1917, killing about 10% of the population. The mill owners had offered a "plague allowance" of 70% for weavers and 20% for spinners. When the epidemic subsided in early 1918, the owners rescinded the bonuses and demanded the workers accept their old pay. The pitifully underpaid workers finally had had enough and began widespread wildcat strikes and a push for unionization.

The mill owners' response to the workers' confrontations was to dismiss their leaders and import outside workers from Bombay. The outcast employees approached the well-known social worker, Anasuya Sarabhai, who was also the sister of the president of the mill owners' association. She subsequently approached Mohandas K. Gandhi for help, who proposed the formation of an arbitration board composed of workers and mill owners. The mill owners at first agreed to participate but soon backed out and locked the weavers out of the factories when the strikes continued. Gandhi responded by aiding the workers in organizing and called for a general strike.

The workers demanded a raise of 50%, Gandhi recommended 35% and the mill owners offered 20%. The mill owners undertook an aggressive pamphlet campaign and the workers began to weaken and return to work. To bolster the sagging workers' resolve, Gandhi began a hunger strike, and within a few days the owners abdicated and agreed to arbitration. They did not have the stomach to make the great soul suffer over such a small affair. To get the mills working as fast as possible, the owners agreed to pay the workers 35% more the first day they returned, 20% the second day and the average of 27.5% for the rest of the days until a decision was made by the arbitrators. Ultimately, the workers were awarded Gandhi's recommendation of 35%.

Boll weevil blues

As if the oppressive sharecropper system was not enough to make southern cotton growers' lives miserable, the boll weevil crossed into the United States from Mexico in 1892 and immediately wreaked havoc on their lives (Yafa 2005; Lange *et al.* 2009). Cotton was still at the core of Southern life and the devastation brought on by the boll weevil was only second to the Civil War. As R.H. Smith (2007) aptly put it – "In the weevils' wake was destruction, devastation, poverty and ruined lives." No other pest before or since has produced such a path of destruction in the United States.

The situation is depicted clearly in the words of the following traditional song (http://ingeb.org/songs/odebollw.html):

> De merchant got half de cotton,
> De boll weevil got de res'.
> Didn't leave de farmer's wife
> But one old cotton dress,
> An' it's full of holes,
> It's full of holes.

The geographical conquest of the weevil through the Cotton Belt was relentless, moving at a rate of 40–160 miles a year. By 1922 it had spread from the eastern two-thirds of Texas and Oklahoma to the Atlantic Ocean. In response, cotton growing shifted to the more isolated western states of California, Arizona and New Mexico, but by the 1980s the weevil had found and devastated even these far-flung areas. Cotton production declined by up to 70% in the hardest hit areas, and total annual losses to the weevil were often above 30% nationally.

For decades, there were few tools known to combat the weevils, and about all the farmers could do was despair. Some relief was gained by burning the woods surrounding the fields to kill weevils hibernating in the litter. It also helped a little to walk through the fields, pinching weevils in the boll or dropping them into kerosene. However, the weevil populations still remained high enough to ravage most farms. Not knowing what else to do, most disbelieving farmers kept planting cotton year after year until the weevil eventually won and they lost their farms.

Some farmers did escape this downward spiral by diversifying from cotton into other crops. Following the suggestions of George Washington Carver at the Tuskegee Institute, peanuts became the alternate crop of choice and rivaled cotton in many areas. Coffee County in Alabama became the peanut capital of the United States in 1917, and the grateful citizens of Enterprise even erected a costly monument to the weevil in the center of their city.

The first effective strategies to control the boll weevil began to emerge in the late 1910s and early 1920s. Farmers began to plant predominantly short-season, early maturing varieties (Yafa 2005). Weevil populations could not grow to such devastating sizes on these early varieties. They also started to dust their fields with calcium arsenate, which killed enough weevils to get profitable yields. The calcium arsenate was applied by hand crank or mule-drawn dust blowers and eventually by plane (crop dusters). It was highly toxic to most animals, including the applicators, but there was really no other alternative known.

After World War II, calcium arsenate was replaced by DDT and other chlorinated hydrocarbons. These were less toxic to humans, but stayed in the environment and animals for years. Rachel Carson's *Silent Spring* chronicled the devastating, long-lasting effects DDT had on the environment. When weevil resistance to the chlorinated hydrocarbons emerged in the 1950s, farmers shifted from organochlorides to organophosphates. These had a much shorter residual time in the environment and were more effective at killing the weevils, but were still highly toxic to warm-blooded animals and birds. To control the weevil, farmers had to flood the environment with chemicals, spraying every three to five days and making 12–16 applications across the season.

Spearheaded by USDA scientists, research began in earnest in the late 1950s to understand the biology of the weevil and develop much safer and effective control strategies (Coppedge and Faust 2003; Kaplan 2003). It was learned in 1959 that late season sprays before the weevils left the undeveloped bolls to hibernate could greatly reduce their populations the following spring. In 1966, scientists discovered that the weevils produced pheromones that attracted mates, and these were artificially produced a year later. Effective, cheap traps were designed in the 1970s that used these pheromones and the first integrated pest management strategies emerged. These used the traps to predict population sizes and time sprays. Pyrethroid chemicals came into use in the 1980s that killed weevils at much lower doses than the previously used organophosphates and were more environmentally friendly.

In 1971, the Agriculture Research Service (ARS) and Animal and Plant Health Inspection Service (APHIS), in partnership with grower associations, began coordinated efforts to eradicate the weevil using the growing arsenal of control strategies. The success of these efforts was truly astonishing. The weevil was essentially eliminated in most of Alabama and Louisiana by 1973, large portions of North Carolina and Virginia by 1979, central Arizona by 1985, and South Carolina, Florida and Georgia by 1897.

The final nail in the weevil's coffin came in 1996 with the release of genetically engineered cotton that carried a modified gene from *Bacillus thuringiensis*, a bacterium that had long been used by organic growers to combat insect pests. The modified gene produces a protein that kills the weevil upon feeding, but is harmless to all other animals and plants except coleopteran insects like the weevil. Today, almost all the cotton grown in the United States and much of the rest of the world contains the engineered Bt gene.

Between the coordinated eradication efforts of the USDA and the use of engineered Bt cotton, weevil populations are now held to manageable levels with only a few chemical sprays. Beneficial organisms that used to be killed unintentionally are now flourishing. The few insecticide sprays are needed mostly to control what used to be secondary pests. The boll weevil that nearly ruined the agricultural economy of the South is now only a nuisance, albeit carefully watched and controlled when necessary.

Garment worker woes and the re-emergence of sweatshops

In the first half of the twentieth century, the number of sweatshops in the United States greatly waned, in large part due to legislation stimulated by the Triangle Shirtwaist fire (Liebhold and Rubinstein 1999). Not only did the legislature of New York respond to the tragedy, but at the federal level Progressives and New Dealers actively worked for reform. During the Franklin Roosevelt administration many programs were established that raised wages, protected the right to unionize and greatly improved the lives of garment workers. These included the National Recovery Administration, the Wagner Act and the Fair Labor Standards Act. In the 1930s, the number of sweatshops in the United States dropped by 26%, and many felt that the sweatshops might actually disappear. An article in *Life* magazine in 1938 essentially declared the sweatshops extinct.

Unfortunately, the disappearance of the sweatshops was ephemeral (Bajaj 2012). As told by Liebhold and Rubenstein in the introduction to *Between a Rock and a Hard Place*:

> In the late 1960s sweatshop production came out of hibernation with a vengeance. A combination of forces at home and abroad contributed to their reappearance. Changes in the retail industry, a growing global economy, increased reliance on contracting, and a new pool of immigrant workers helped to bring about the return of American sweatshops. (1999, p. 7)

Sweatshops also cropped up in Latin America, China and Asia where garments could be made cheaply due to the tiny wages paid, as low as ten or twelve cents an hour.

Several factors led to the re-emergence of the sweatshops. The structure of the retail industry changed radically from small local shops to national chains and giant retailers who had their own private labels and contracted their garments out for manufacture. Consumer demands for lower prices led contractors to pressure manufacturers to reduce wages and provide a cheaper product. This pressure for lower-priced goods also encouraged the garment business to begin migrating to the non-union US South and Southwest, and across the developing world. The emergence of computers in the 1980s also allowed retailers to reorder garments as they were sold, rather than needing to maintain huge inventories. This made garment workers' already volatile work hours even more unpredictable and increased the pressure on them to produce more in less time.

In the United States, the most notorious sweatshops arose in Manhattan's Chinatown and El Monte, California. The Chinatown sweatshops began in the 1940s and reached their peak in the 1980s, preying on thousands of Chinese immigrants attracted to the cheap rents in the area. The workers toiled in converted tenement apartments for twelve to fifteen hours a day, seven days a week with no minimum wage. The workers were paid a piece rate, as low as eight cents a garment, resulting in average wages of about a dollar an hour. The rooms became hot and stifling from the heat of the sewing machines as there was no air conditioning. Bathroom facilities were generally filthy and the toilets often did not flush. Many children helped their mothers as best they could.

As bad as these conditions sound, it was worse for garment workers in El Monte, California in the family-run business of Suni Manasurangkun. Poor people from rural Thailand were recruited by promises of higher wages in a clean factory, with weekends off and trips to Disneyland. They were enticed to sign agreements that indentured them for three years and required a payback of $5,000 (US dollars in 1997). The Thais were smuggled into the United States with fake passports and immediately upon arrival were locked up in a factory complex, where windows were boarded up and razor wire surrounded the complex to prevent escape. The poor, disoriented Thais were forced to toil from dawn to midnight in overheated, stifling rooms; they collapsed at night in mattress-filled rooms in groups of ten and were allowed little contact with the outside world. For their efforts, they were paid a pitiful average of about sixty-nine cents an hour and most of that money went to pay for their food and their payback. Armed guards roamed the property and the workers were constantly screamed at to work harder and faster.

This operation lasted almost a decade before officials finally busted it. The first attempt came in 1988 when a disgruntled competitor blew the whistle and complained about the salaries being paid. Unfortunately, the Immigration and Naturalization Service (INS), which had been alerted, served a notice of inspection before their search and the workers were moved before the authorities got there. Another planned raid in 1992 by the INS was thwarted by high officials. Finally, in 1995, the sweatshop was successfully raided after the California Department of Industrial Relations was tipped off by the boyfriend of an escaped worker (Anonymous 1998). Ten operators were indicted, including Suni Manasurangkun and her sons. They received prison sentences of 6–7 years and required to pay fines of up to $4.5 million to the migrant workers. Eleven companies that had contracted with the El Monte sweatshop also agree to pay $3.7 million to the 150 workers who had worked there. These companies did not admit any guilt, stating that they did not know the operators were abusing the workers and breaking the law.

The public outcry associated with the El Monte sweatshop resulted in a firestorm of private and governmental initiatives. President Clinton in 1996 formed the White House Apparel Industry Partnership to search for solutions. Congress significantly increased the number of Department of Labor wage and hour investigators. Numerous activist groups began media campaigns against businesses that used sweatshops. These efforts have improved conditions but there is still a long way to go. Liebhold and Rubenstein reported that "in 1996 the US Department of Labor estimated that out of 22,000 US garment shops, at least half were in serious violation of wage and safety laws."

Globalization of the textile industry

In the 1980s, the whole structure of the retail industry in the United States began to shift dramatically (Liebhold and Rubinstein 1999; Anonymous 2014b). As was previously described, the former bedrock of the industry, small local shops, began to rapidly disappear in favor of national chains and giant retailers with private labels. Prices began to drop as the retailers competed fiercely for consumers. Contractors and manufactures were pressured to provide cheaper and cheaper goods. This encouraged a reduction in wages and the movement of the industry to the non-union South and

Southwest. By 1995, there were only 355,000 unionized textile workers left in the United States, down from 850,000 in 1970.

In the search for ever-cheaper products, the US retailers eventually started buying a significant portion of their clothing from offshore manufactures in far-off places like China, Korea, Bangladesh, Cambodia, Pakistan and Honduras. These suppliers paid salaries a fraction of those in the United States. The national chains and wholesalers turned a blind eye to how the manufacturers treated their workers, as long as they could get huge quantities of cheap products. Offshore garment workers were paid as little as thirty cents an hour, in conditions almost rivaling those of the late 1800s and early 1900s in the United States and Great Britain. Many of the factories were death traps, crowded and almost tinderboxes, subject to fire with no sprinklers and few exits. One of the most tragic events occurred in Bangladesh in 2012, when over 100 people died of smoke inhalation and burns, and hundreds more were injured. In fact, more than 500 Bangladeshi workers have died in factory fires in the last decade.

Numerous media campaigns were begun by activists against those retailers that bought their textiles from these global sweatshops. Considerable heat was brought onto companies such as Nike, Walmart, The Gap, Tommy Hilfiger, Calvin Klein, Disney and others. Heat was also brought upon celebrities such as Kathy Lee Gifford and Michael Jordan, who have signature lines with these companies. Because of the public outcry, US retailers have begun to more carefully track worker conditions at their suppliers' factories and buy more goods from US suppliers; however, real change is still only in its infancy. Most companies rely on voluntary auditing and inspection, and are reluctant to contribute directly to plant improvements and fire safety.

Cotton and textile industry today

World cotton use is growing at about 2% a year and is expected to continue at that rate for the next decade (OECD-FAO Agricultural Outlook 2014). The United States is still the world's leading exporter of cotton products, with about 24% of the trade; India is second with about 18%. World cotton production is growing to match the demand. India and China are the largest producers, followed by Bangladesh, Vietnam, Turkey, Pakistan and Indonesia.

China traditionally has produced the most cotton in the world, but India is now hot on its heels. The cropping practices in China are labor intensive and profits have been stifled by steadily rising wages. Fragmented landholding in China has made mechanization difficult, particularly in the eastern provinces.

More than half of the cotton grown across the world is genetically engineered for herbicide and insect resistance (Anonymous 2015). The leading countries are the United States (almost 100% adoption), India (> 90%) and China (> 60%). Significant GE cotton acreages can also be found in

Argentina, Australia, Mexico, Brazil, Colombia, South Africa, Burkina Faso and Costa Rica.

In the last few years, the textile industry in the United States has rebounded, particularly in the South (Anonymous 2014b). Hundreds of thousands of jobs were lost to cheaper labor offshore and automation in the 1990s and 2000s, and over 600 plants closed. New plants are once again being opened and existing ones expanded due to rising wages in China, increases in transportation costs and tariffs. Even Walmart has pledged to buy $50 billion worth of US textiles over the next ten years. International companies are investing in US plants including Brazil, Canada, China, Dubai, United Kingdom, India, Israel, Korea, Mexico and Japan. Many states are offering grants to encourage job growth.

The biggest growth in US textiles has been in non-woven, compressed fabrics such as diapers, wipes and mops. Today's textile plants are very different from those of the past and rely heavily on automation. Far fewer people are employed in the factories and the workers of today must be more technology savvy. As a result, average wages have risen 15% to 20% in the last couple of years, although the average textile worker's salary is still poor compared to other Industries, averaging $40,000 a year, compared to $60,000 in other manufacturing jobs.

References

Anonymous (1998) 4 draw prison terms in sweatshop case, *NY Times* (May 8).

Anonymous (2011) *The 1911 Triangle factory fire*, Kheel Center, Cornell University Library, http://trianglefire.ilr.cornell.edu/story/introduction.html.

Anonymous (2014a) *Mill life in Lowell: 1820 to 1880*, Center for Lowell History – University of Massachusetts Lowell Libraries, http://library.uml.edu/clh/mo.htm

Anonymous (2014b) Textile industry comes back to life, especially in South, *USA Today* (Feb. 5)

Anonymous (2015) Genetically modified cotton, *GMO Compass*, http://www.gmo-compass.org/eng/database/plants/21.cotton.html.

Bajaj, V. (2012) Fatal fire in Bangladesh highlights the dangers facing garment workers, *NY Times* (Nov. 26)

Batchelder, S. (1863) *Introduction and early progress of the cotton manufacture in the United States*, Little, Brown, Boston, MA.

Berdan, F.F. (1987) Cotton in Aztec Mexico: Production, distribution and uses. *Mexican Studies/Estudios Mexicanos*, vol. 3, 235–262.

Bordewich, F.M. (2012) *America's great debate: Henry Clay, Stephen A. Douglas and the compromise that preserved the Union*, Simon and Schuster Paperbacks, New York.

Brite, E.B. and Marston, J.M. (2013) Environmental change, agricultural innovation, and the spread of cotton agriculture in the Old World. *Journal of Anthropological Archaeology*, vol. 32, 39–53.

Brubacker, C.L., Borland, E.M. and Wendel, J.E. (1999) The origin and domestication of cotton. In: *Cotton: Origin, history, technology and production* (pp. 3–31). Smith, C. W and Cothren, J.T. (eds.). John Wiley and Sons, New York.

Choi, N. (2012) *Ahmedabad textile laborers win strike for economic justice, 1918*, Global Nonviolent Action Database, http://nvdatabase.swarthmore.edu.

Copeland, M.T. (1917) *The cotton manufacturing industry of the USA*. Harvard University Press, Cambridge, MA.

Coppedge, J. and Faust, R.M. (2003) Winning the weevil war: Beating a $22 billion bug. *Agricultural Research*, vol. 51, 2.

Davis, D.B. (2006) *Inhuman bondage: The rise and fall of slavery in the New World*, Oxford University Press, New York.

Fishman, M., Causey, V. and Woodrum, R. (2014) *Georgia State University Library*, Special Collections and Archives: Work n' Progress – Lessons and Stories, http://research.library.gsu.edu.

Fite, E.D. (1910) *Social and industrial conditions in the North during the civil war*, Macmillan, New York.

Hall, J.D., Leloudis, J., Korstad, R., Murphy, M., Jones, L.A. and Daly, C. (1987) *Like a family: The making of a southern cotton mill world*, University of North Carolina Press, Chapel Hill.

Hebron, J. (1988) *The emergence of the cotton kingdom in the old Southwest: Mississippi, 1770–1860*, Louisiana University Press, Baton Rouge.

Hindman, H.D. (2002) *Child labor: An American history*. M.E. Sharpe, Armonk, NY.

Hutchinson, J.B. (1951) New evidence on the origin of the old world cottons. *Heredity*, vol. 8, 225–241.

Hutchinson, J.B., Silow, R.A. and Stevens, S.G. (1947) *The evolution of Gossypium*, Oxford University Press, London.

Johnson, W. (2013) *River of dark dreams, Slavery and empire in the cotton kingdom*, Belknap Press of Harvard University, Cambridge, MA.

Kaplan, J.K. (2003) We don't cotton to the boll weevil around here no more. *Agricultural Research*, vol. 51, 4–8.

Kingsley, C. (1850) *Cheap clothes and nasty*, W. Pickering, London.

Lange, F.A., Olmstead, A.L. and Rhode, P.W. (2009) The impact of the boll weevil, 1892–1932. *Journal of Economic History*, vol. 69, 685–718.

Leloudis, J. and Walbert, K. (2014) *Like a family*, Website developed by the American Historical Association's program Teaching and Learning in the Digital Age, funded by the National Endowment for the Humanities, http://www.ibiblio.org/sohp/laf/overview.html.

Liebhold, P. and Rubinstein, H. (1999) *Between a rock and a hard place: A history of American sweatshops 1820 – Present*, UCLA Asian-American Studies Center and Simon Wiesenthal Center Museum of Tolerance, Los Angeles, CA.

May, O.L. and Lege, K.E. (1999) Development of the world cotton Industry. In: *Cotton: Origin, history, technology and production* (pp. 65–97). Smith, C.W. and Cothren, J.T. (eds.). John Wiley and Sons, New York.

Mitchell, C.C. (2008) The development of cotton from the Old World to Alabama: Chronological highlights in Alabama cotton production. *Agronomy and Soils*, Departmental Series No. 286. Alabama Agricultural Experiment Station, Auburn University, Alabama.

Moore, J.H. (1956) Cotton breeding in the old south. *Agricultural History*, vol. 30, 95–104.

Northup, S. (2014) *12 Years a slave*, First edition – Derby and Miller, 1853; current edition Graymalkin Media, Los Angeles, CA.

O'Brien, P., Griffiths, T. and Hunt, P. (1991) Political components of the Industrial Revolution: Parliament and the English cotton textile industry, 1660–1774. *Economic History Review*, vol. 44, 395–423.

OECD/Food and Agriculture Organization of the United Nations (2014) *OECD-FAO agricultural outlook 2014*, OECD, http://dx.doi.org/10.1787/agr_outlook-2014-en.

Parakunnel, J.T. (1924) The beginnings of calico-printing in England. *English Historical Review*, vol. 39, 206–216.

Phillips, L.L. (1979) Cotton. In: *Evolution of crop species*. Simmonds, N.E. (ed.). Longman, London.

Riello, G. (2010) Asian knowledge and the development of calico printing in Europe in the seventeenth and eighteenth centuries. *Journal of Global History*, vol. 5, 1–28.

Riis, J.A. (1890) *How the other half lives: Studies among the tenements of New York*, Charles Scribner's Sons, New York.

Rosenberg, C.M. (2010) *The life and times of Francis Cabot Lowell, 1778–1817*, Lexington Books, Plymouth, UK.

Schlingloff, D. (1974) Cotton manufacturing in ancient India. *Journal of the Economic and Social History of the Orient*, vol. 17, 81–90.

Smith, R.H. (2007) *History of the boll weevil in Alabama, 1910–2007*, Alabama Agricultural Experiment Station Bulletin 670, Auburn University, Auburn, AL.

Stephens, S.G. (1967) Evolution under domestication of the New World cottons (*Gossypium* spp.). *Ciencia Cultura*, vol. 19, 118–134.

Stephens, S.G. (1976a). Some observations on photoperiodism and the development of annual forms of domesticated cottons. *Economic Botany*, vol. 30, 409–418.

Stephens, S.G. (1976b) The origin of sea Island cotton. *Agricultural History*, vol. 50, 391–399.

Sujata, P. (1984) Class conflict and workers' movement in Ahmedabad textile industry, 1918–23. *Economic and Political Weekly*, vol. 19, 853–864.

Wendel, J.F. (1989) New world tetraploid cottons contain old world cytoplasm. *Proceedings of the National Academy of Sciences USA*, vol. 86, 4132–4136.

Wendel, J.F., Brubaker, C.L. and Seelanan, T. (2010) The origin and evolution of *Gossypium*. In: *Physiology of cotton* (pp. 1–18). Stewart, J.M., Oosterhuis, D.M., Heitholt, J.J. and Mauney, J.R. (eds.). Springer Science+Business Media, New York.

Wendel, J.F., Olson, P.D. and Stewart, J.M. (1989) Genetic diversity, introgression, and independent domestication of old world cultivated cottons. *American Journal of Botany*, vol. 76, 1795–1806.

Yafa, S. (2005) *Cotton: The biography of a revolutionary fiber*, Penguin Books, New York.

4 Tea

Sources of tea

There are four major types of tea consumed in the world – green, black, oolong and white. All of these come from leaves plucked from two varieties of *Camellia sinensis* – var. *sinensis* (tea) and var. *assamica* (Assam tea) (Chen *et al.* 2012; Mondal 2014). The var. *sinensis* is native to China and is a small-leaved shrub that is 1–5 meters in height. The var. *assamica* is native to India and is a large-leaved tree that reaches 8–12 meters in height.

There are a number of "infusions" of leaves that are commonly called "tea," but are not botanically true teas (Martin 2007). These herbal or tisane teas include *rooibos* from South Africa, *yerba maté* from South America, mint teas from Morocco and the traditional herbal and fruit "teas" of Turkey, which are now more prevalent across Europe and internationally.

The various kinds of teas are produced by different manufacturing methods. The fresh leaves for green tea are treated at high temperatures to prevent their oxidation before drying. The green teas have low levels of caffeine and have subtle flavors. For black tea, the leaves are allowed to wither and oxidize before drying. This produces a characteristic dark brown to black leaf with a robust flavor and high caffeine content. Oolong tea is produced by allowing the leaves to wither and ferment for short periods of time before being heat-treated to stop the oxidation process. This yields a tea with a very fragrant taste and an aroma likened to fresh flowers and fruit. White teas are produced by selecting only the youngest leaves and preventing oxidation. These are delicate, complex teas with a natural sweetness.

All the tea in China

No one really knows when humans started making tea, but it is likely that early Cro-Magnon were chewing raw leaves for thousands of years before the advent of agriculture. The caffeine found in the leaves would have given them a slight buzz and increased alertness. Even today, the people who live near wild stands of tea still chew the raw leaf and ferment boiled leaves for food.

Chinese legend has it that the first cup of tea was made by the second emperor of China, Shen Nung, about 5,000 years ago (Standage 2005; Mair and Hoh 2009; Chen, Apostolides and Chen 2012). Nung is considered to be the father of Chinese medicine and pharmacology, and wrote the treatise *On Herbal Medical Experiment Poisons*. He personally tasted hundreds of herbs to study their effects and actually died of a toxic overdose. The story goes that the first cup of tea was accidently brewed by Nung when a gust of wind blew a few leaves into water he was boiling over wood from a tea plant. An experimentalist to his core, Nung is said to have tasted the brew and found it to be both refreshing and stimulating.

Whether or not this story is true, tea drinking didn't take off across China until the last century BC. This is the date when the first reference to the boiling of tea appears in the Chinese literature. Tea was restricted to royalty for centuries before it became a drink of the common man. During this period, fresh tea leaves were steamed and processed into cakes, called cake-tea.

During the Eastern Han dynasty (AD 25–200), tea cultivation and consumption gradually spread down the Yangtze River across the center of China nearly to Shanghai. The Buddhist monks began drinking tea in the fourth century and further spread its popularity. There were over 40,000 monasteries scattered across China housing millions of monks. Monks cultivated tea, drank it during meditation and study, sold it, gave it to visitors and offered it in tribute to Buddha and the royal court. Wandering far and wide was an integral part of a monk's life, serving as ambassadors of cultural life (and tea!) to royalty and peasants alike.

Tea had become the most popular drink in all of China by the Tang dynasty (618–907), replacing rice and millet beer. It became so popular that the government began taxing its sale in 780. In that same year, the famous Taoist poet, Lu Yu, published *The Classic of Tea*. In this seminal piece, Yu went into great detail on how tea was grown, prepared and served, and elevated tea from a simple pleasure to a drink of high social standing (Martin 2007). As Tom Standage describes in his book, *A History of the World in 6 Glasses*:

> Lu transformed tea from a mere thirst-quenching drink to a symbol of culture and sophistication. Tea tasting and appreciation, particularly the ability to recognize different types, became highly regarded. Making tea became an honor reserved for the head of the household, an inability to make tea well, was considered a disgrace. Drinking parties and banquets centered on tea became popular at court, where the emperor drank special teas made with water transported from particular springs. (p. 181)

The Tang dynasty emperor, Dai Zong (762–779), planted the first imperial tea garden in 771 on Guangzhou Mountain near Yixing, east of Shanghai. The plantation grew rapidly and by 785 required 30,000 laborers to pick and process the leaves. These unfortunate workers were pulled from their own fields, causing them much misery and inconvenience. The tea from this field was rushed by tea horse caravan each year to the Royal Court in Chang'an, just in time for the annual Qingming Festival, honoring the dead in China.

In the tenth century, China had a little Ice Age, when average temperatures fell about 5°F. The cold killed many plants at the Guangzhou Mountain tea estate and the development of the survivors was delayed, making the harvest too late for the Qingming Festival. In 977, the imperial tea estate was moved to the Fujian province in southeast China, where temperatures were much milder. An immense estate was developed there, spreading five miles along the eastern branch of the Jian River. The royal property came to contain twenty-five separate tea gardens and thirty processing facilities. Again, thousands of local people were called from the adjacent hillsides to pluck leaves during the harvest season.

During the Song dynasty (960–1279), modes of tea preparation became quite ritualized. Teas also became known for the regions they were grown, the water used to boil them and how the tea leaves had been processed. Tea drinking was now a way of life for both royalty and common folk.

Tea use in China came to a crashing halt in the thirteenth century when the Mongol horsemen from the North took control of the country. At the empire's peak, Genghis Khan ruled a vast area from the middle of Europe to Korea and south to Vietnam. The Mongol's drink of choice was koumiss, an alcoholic concoction made by churning and fermenting mare's milk in a leather bag. While it is hard to believe that all common folk in China stopped drinking tea, it fell out of favor among the elite. Tea drinking roared back when the Mongols lost control of China and the Ming dynasty was established in 1368. Tea quickly returned to its former status, not only as a popular refreshment, but also as a spiritual drink offering contentment, peace and hope.

Up until the late 1500s, only green tea was produced in China. To maintain the green color, the leaves were typically heated after plucking. The problem with green tea was that it could only be stored for short periods of time, and could therefore not be traded long distances. To circumvent this problem, the Chinese learned to let the leaves wither without heating, producing a more potent, longer storing "black" tea. They also learned to make "oolong teas," where the harvested leaves were bruised by shaking before fermentation.

On to Japan

Tea was first brought to Japan in the first century AD by Buddhists (Standage 2005; Mair and Hoh 2009). The Japanese monks Saichō and Kūkai traveled to China in 804 to study Buddhist texts, and while there became enamored with tea and the ceremonies surrounding it. Upon his return in 805, Saichō carried tea seeds back with him and planted the first Japanese tea garden in a shrine near the capitol of Kyoto at the base of Hiei Mountain. Unfortunately, Saichō's efforts came to little, as tea drinking did not take off in Japan for almost another 400 years.

The person responsible for making tea popular in Japan was the Buddhist monk, Myōan Eisai. Like Saichō and Kūkai, he learned to love tea while studying in China and brought seeds back with him in 1191 and planted them at two locations – at the Sankōji Temple on the island of Hirado near Nagasaki, and at Tendaijiin on the island of Kyushu. He also sent seeds to the abbot Kōzanji in Kyoto, who had them planted in the mountains of Togano and at the city of Uji (north and south of Kyoto, respectively). The plants did well at all locations, but those set at Uji produced exceptional tea, becoming the benchmark for all tea produced in the future.

Eisai went to China to become a Zen master and returned with an almost mythical belief in the healing properties of tea. Japan had been through a tumultuous period of warfare, natural disasters and famine, and Eisai was convinced that the teachings of Zen Buddhism could restore the country to its past glories. He also believed that the restorative properties of tea could play an important role in that rebirth. In 1214, he found a strong supporter of his beliefs when he sobered up the Shogun Minamoto Sanetomo with a cup of tea and then taught him about the health benefits of tea. Through Eisai's and Sanetomo's advocacy, the popularity of tea spread far and wide across China. Even the humblest families started to plant a couple of tea bushes around their homes.

In the sixteenth century, Sen-no Rikyu, the great tea scholar, Buddhist and tea master to the emperor, promoted the drinking of powdered green tea, which became the drink of choice among the samurai and aristocracy. The type of green tea drunk today was developed by Sohen Nagatani in the eighteenth century. In his process, the young leaves were harvested, steamed rather than roasted, kneaded and then dried.

As the popularity of tea grew by leaps and bounds in Japan, so did the intricacies of the Japanese tea ceremony. All aspects of what became almost a spiritual event were choreographed with rigid rules and timing (Martin 2007). As Tom Standage describes it:

> The full Japanese tea ceremony is an immensely intricate, almost mystical ritual that can take more than an hour. Merely to describe the steps of grinding the tea, boiling the water, mixing and stirring the tea is to overlook the significance of the particular form of utensils, and the order and nature of their use. The water must be transferred from a specific kind of jar to the kettle using a delicate bamboo dipper; a special spoon is used to measure out the tea; there must be a special stirrer, a square silk cloth to wipe the jar and spoon, a rest for the kettle lid, and so on. All these items are to be brought in by the host in the correct sequence and placed on the correct mats. Ideally, the host is even to

gather the firewood himself, and the whole ceremony should take place in a teahouse situated in an appropriately laid-out garden.

(pp. 183–184)

China tea to Tibet and Russia

In 1578 Tibetan religious leaders joined together with the Mongolian rulers to request trade with China for tea. Buddhist monks had enjoyed tea for centuries and it was no longer forbidden in Mongolia (Martin 2007; Mair and Hoh 2009). When the Chinese authorities agreed, it didn't take long for tea to become an integral part of all Tibetan and Mongolian life. Monasteries began serving tea eight to ten times a day.

The trade route to Tibet was arduous to say the least. In the late 1500s, it started in Kalgan, 100 miles northeast of Peking, and climbed to Xining in western China on the Tibetan Plateau. This city, at over 7,000 feet in elevation, was an important commercial center on the Northern Silk Road for over 2,000 years. From Xining the route continued over towering mountains to the Himalayan Plateau, finally arriving in the Tibetan city of Lhasa at an altitude of almost 11,500 feet. In the late sixteenth century, the trade route shifted to Dajianlu (now Kangding), at the historical border between China and Tibet. Porters carried tea and other goods from Yazhou, in Sichuan Province, to Dajianlu, crossing over a 370-foot suspension bridge that spanned the Dadu River at Luding.

The tea going to Tibet was mostly black brick tea, varying greatly in quality. The tea was chopped and steamed, dried and then packed in bamboo cylinders holding twenty to twenty-five pounds each. These were loaded on the backs of porters who carried astonishing loads of 300 pounds or more. The men trudged along at a rate of about six miles a day, finishing with a final eighteen-mile-long, 3,000-foot ascent to Dajianlu. Here the tea was repackaged into hides and carried by yaks to Lhasa, the terminus of a 1,300-mile journey that took about three months.

Figure 4.1 Men carrying tea bricks for Tibet

Photographer: E.H. Wilson, 1908. Source: wikimedia.org.

The tea route to Russia was just as arduous as that from China to Tibet. By treaty in 1727, Kyakhta near the Mongolia–Russia border became the crossing point for trade between China and Russia. Similar to what happened in Tibet, once free trade in tea was allowed, it rapidly became the non-alcoholic drink of choice in Russia. Unique rituals and utensils developed around its consumption, and elaborately decorated samovars for the boiling of tea became the centerpiece of Russian homes.

The caravan route from Kalgan to Kyakhta covered over 1,500 miles across the desolate Mongolian steppe. During the hot summer months, when forage was plentiful, caravans of several hundred oxcarts carrying 400-pound loads traversed the route. In the winter months, camels were the favored beasts of burden, loaded with 280-pound baskets of tea. In 1818, 1,420 oxcarts and 3,450 camels crossed the route, and by 1829 the route was covered by 2,705 oxcarts and 9,670 camels.

Once in Kyakhta, the Russians moved the tea by wagons, horse sleighs and barges through southern Siberia and over the Ural Mountains to the distant trading emporium, Nizhny Novgorod along the banks of the Volga River (250 miles east of Moscow). Half of all the goods traded in Russia came through this market. When the tea had finally arrived at Nizhny Novgorod, it had covered an astonishing 3,800 miles, in a trip that took 100 to 200 days.

Tea arrives in Europe

When Europeans first started to knock on the door of China in the sixteenth century, China considered itself a "Celestial Empire" in need of nothing from no one (Standage 2005; Mair and Hoh 2009; Rose 2009). The Chinese actually held the rest of the world in contempt, believing them to be barbarians. China was by far the largest country and it had the most people. Little of the technology found in Europe had not been first invented in China.

But the Chinese did want gold and silver, something the Western world could provide. To get these precious metals, the Portuguese were allowed to establish a trading post at Macao in the Canton estuary in 1557. Trade was limited to this single location and was the sole purview of the Portuguese. With this rigid control of trade, the Chinese could restrict contact with the outside world and levy duties they knew would be collected.

While the Portuguese traders certainly had the opportunity to buy tea in China and ship it to Europe, they focused mostly on spices and silk. The Dutch East India Company (DEIC) became the primary supplier of tea to Europe, sending their first shipment in 1610. The DEIC shipped tea to Europe from its intermediaries in Japan, Taiwan and mainland China through a trading hub it established in Batavia (now Jakarta), Indonesia.

By the 1630s, tea had become the drink of fashionable people in Dutch society. It was simply too expensive for the common person, probably costing well over a thousand dollars a pound in today's dollars. The first tea was also being sold in England at about the same time, but only in pharmacies as medicine. It wasn't until 1657 that the first teahouse opened there.

In the mid-seventeenth century, most of the tea in Europe came from Japan via the Dutch East India Company. They gained a monopoly on trade from Japan in 1638, when the Portuguese were thrown out of the country for pushing Christianity too hard. Jesuit missionaries had converted hundreds of thousands of Japanese by the second half of the sixteenth century, and the rulers had become increasingly alarmed by what they considered an external threat. This left the door open to the DEIC, who wound up with a stranglehold on Japanese trade that lasted until the DEIC was dissolved in 1795 after the Fourth Anglo-Dutch War.

The English East India Company (EIC) began to bring tea into England from China in 1689. Their trade was centered in Canton, where the Chinese had allowed English and French "foreign devils" to establish factories. The amount of tea moved by the EIC was initially modest – in 1678 they shipped only about 5,000 pounds to England. The company's highest profits came from Indian cottons brought to England and France to feed the "chintz craze." The EIC didn't really set its sights on tea until the Industrial Revolution made England the world's leading textile manufacturer, and the more expensive Indian cottons fell out of favor.

The shift from cotton to tea turned the EIC into a commercial powerhouse. As Mahr and Hoh put it in *The True History of Tea*:

> The EEIC's switch to tea proved fortuitous, for through this trade, it became the wealthiest, most powerful commercial enterprise in the world, able to command armies, challenge nations, make war, form alliances, issue its own money and run its own courts of justice.

At the peak of its power, 60% of the EIC trade was in tea and 10% of the British government's revenue came from duties on tea.

English become obsessed with tea

The first person to incorporate tea drinking into British life was Catherine of Braganza, who married Charles II in 1662 (Standage 2005; Martin 2007; Mair and Hoh 2009). She began entertaining with tea served in thimble sized porcelain cups, a practice quickly embraced by the aristocracy. Her father, King John IV of Portugal, had given as her dowry two trading posts in Tangier and Bombay, along with the right to trade with the Portuguese ports spanning Africa, India and southeast Asia.

Queen Anne (1702–1714) was the first of the British royalty to make tea a regular part of the day. She scandalized the court by replacing ale with tea at breakfast, also serving it in tiny porcelain cups. This was probably not enough tea each morning to create a buzz, but its consumption was certainly more stimulating than ale! Queen Anne is also credited with instituting the mid-afternoon tea break with cake, bread and butter. In addition, during her reign, the British began to add sugar to their tea, supporting the burgeoning British sugar industry in the West Indies.

Tea did not catch on with common people until the end of the seventeenth century. It was simply too expensive until then. In 1660, a pound of the best teas was selling for £4–10 and even as late as 1700, a pound still cost about £4. This was well out of reach for a poor, working family whose earnings were perhaps £20 a year.

Tea consumption was also greatly eclipsed by coffee drinking among the middle class during this time. Coffeehouses were all the rage in England by the late 1600s, serving as centers of knowledge where men could come and discuss politics, ideas and schemes. Women were excluded. In the early 1700s there were 2,000 coffeehouses in England. Some of these coffeehouses also offered tea, but the cost of a cup of coffee was five times cheaper.

The first exclusive teahouse in London was opened in 1717 by Thomas Twining. He took the bold step of allowing women into the establishment to drink and buy his blended teas. Suddenly there was a place for both sexes of genteel people to mingle and chat. Another innovation, tea gardens, began to pop up all across England in 1732. These were essentially open-air restaurants where respectable couples could stroll along flowered walkways under decorated arbors. Many of the tea gardens had bandstands, concerts, entertainers, gambling and fireworks.

Tea drinking got a huge boost in popularity at the end of the eighteenth century, as a by-product of the Industrial Revolution. To provide a refreshing break and keep minds sharp, factory workers were encouraged to drink tea. The companies even provided free tea breaks. Workers were much more productive and safe consuming tea rather than the previous drink of choice, beer. Regular tea drinking also greatly reduced dysentery among workers. The water was purified by boiling and tea's tannin content was a natural deterrent to disease.

By the end of the eighteenth century, coffee use had dropped precipitously and almost everyone in England was drinking tea. Between 1699 and 1799, annual legal tea imports had risen dramatically from about six tons to 11,000 tons. Smuggling was also rampant, likely doubling the amount of tea finding its way into England. This massive amount of importation brought the price of tea down dramatically, falling by 95% from the beginning to the end of the eighteenth century. The huge amounts of tea traded by the EEIC had simply overwhelmed coffee imports.

Clipper ships

As the demand for tea grew, so did the importance of getting the tea to London as fast as possible from China (Martin 2007; Mair and Hoh 2009). The first ships to arrive in port with the new crop received the highest prices as people clamored for the new supply of tea. The first ships of the East India

Company were ponderous and slow, and were soon replaced by the sleek "clipper ships" developed by the Americans during the War of 1812. The first clipper ship to travel from China to London was three times faster than those of the East India Company and made the trip in a little more than three months. In the heyday of the clipper ships in the 1860s, up to forty ships at a time would race from China to London and people would line the docks waiting to see which would be the first to arrive. The development of efficient, fast steamships and the opening of the Suez Canal in 1869 brought an end to the glory days of the clipper ship.

Tea and revolution

For the first half of the 1700s, Britain pretty much left the colonies alone, allowing them to trade at will and taxing them far less than other British subjects (Standage 2005; Mair and Hoh 2009; Philbrick 2013). The colonies were by law supposed to trade only with Great Britain, but in fact illicit smuggling was rampant. Many merchants had grown wealthy shipping molasses from the French West Indies into America to make rum, and up to 75% of the tea consumed by the colonists came from St. Eustatius in the Dutch West Indies. The first three American millionaires, T.H. Perkins, Stephan Girard and John Jacob Astor, made their fortunes by bypassing the East India Company and shipping tea directly from China (Martin 2007).

As was discussed in the chapter on sugar, this privileged situation abruptly changed when the British Parliament decided in 1765 that it needed to recoup the costs Britain had incurred in America during the French and Indian Wars. It had accumulated a debt of $22.4 billion in today's dollars, and Parliament wanted the colonists to help foot the bill. To do this, they passed the Stamp Act, which would require a revenue stamp on all printed materials.

The colonists' reaction was outrage. They had grown quite comfortable running their own affairs with minimal interference from Britain. Now they were being directly manipulated with a tax levied on them by a distant Parliament without representation. Huge protests erupted on the streets of Boston and one mob even ransacked the home of the lieutenant governor, Thomas Hutchinson. A shocked British Parliament repealed the Stamp Act in 1766, but replaced it a year later with the Townshend Revenue Act, which levied taxes on paper, lead, glass and tea. This revenue was intended to pay the salaries of the British appointed judges and governors sent to the colonies.

The British government thought that these taxes would be more palatable to the Americans than the Stamp Act, because the revenue would go to pay colonial officials. However, the colonists saw the Townshend Acts as another provocation. First of all, they were being taxed again without representation and second, they would lose the little control they currently had over their judges and governors. The colonial legislators had been raising local taxes to support these officials, making the officials somewhat beholden to them.

Now they would have no influence over them at all, as their salaries would be coming from the British treasury.

Another part of the Townshend Revenue Act that grated on the colonists was that it established an American Customs Board of five commissioners. These commissioners were charged with eliminating smuggling and were given permission to search homes, shops and ships without warrant. Colonial merchants like John Hancock were horrified, as most of them had garnered their wealth through smuggling.

The colonists responded to the Townshend Revenue Act by boycotting all British goods and protesting in the streets. A riot ensued when one of John Hancock's ships, the *Liberty*, was searched and seized. This time the British government decided to exert control and sent several regiments of soldiers to Boston to maintain the peace. Not surprisingly, the colonists were infuriated by what they considered to be an occupation by the British. Their unhappiness reached a fever pitch when on March 5, 1770, a handful of soldiers got trapped in the middle of an angry, taunting crowd, panicked and fired their weapons indiscriminately, killing five. The streets overflowed with livid protestors reacting to what became known as the "Boston Massacre."

A town meeting was held where demands were made for the removal of the British and a murder trial for the soldiers. The British authorities in America decided to defuse the situation by withdrawing the British Regulars from Boston proper. The British Parliament also repealed most of the Townshend Acts except the tax on tea, to placate the colonists. Colonial lawyers John Adams and Josiah Quincy defended the British in a murder trial, and got them acquitted on a plea of self-defense.

Unfortunately, the British Parliament could not leave well enough alone, and in 1773 passed the Tea Act, in an attempt to bail out the debt ridden East India Company. The smuggling of tea into the colonies had resulted in huge stockpiles of tea in London warehouses, estimated at 10,000 tons. The company was required to pay import duty on that tea, regardless of whether it was sold or not, and as a result they owed back taxes of over a million pounds.

In the Tea Act, the Parliament gave the EIC a loan of £1.4 million and the right to ship tea directly to America from China and therefore skip the British import tax. This meant they could sell the tea much cheaper in the colonies, in the hopes they could undercut the smugglers. In addition, the Act established that "company agents" would be identified in America who would have exclusive rights to collect those taxes. The Parliament felt that the EIC could not be allowed to fail, as it was responsible for too much of the British economy and had too many friends in high places.

The misguided British believed that this time the colonies would be happy with a law, because it would get them cheaper tea and it did not impose any new taxes on them. This was, of course, wishful thinking, as the firebrands of the revolution saw it as another attempt to force a British commercial monopoly on the colonies. The seeds of revolution had by now been irrevocably sown.

When the first three ships loaded with tea reached Boston Harbor in 1773, the colonists prevented them from being unloaded and demanded that they be sent back. The now royal governor Hutchinson refused to let them leave, further stirring the pot of resistance. On the night of December 16, 1773, in what came to be called the "Boston Tea Party," 116 colonists dressed as Mohawk Indians boarded the three ships and dumped 342 chests of East India Company tea into the harbor. The tea weighed over 92,000 pounds and was valued at more than $1.7 million in today's dollars. For the next several days, the harbor stunk of tea.

The enraged British Parliament responded by passing the first of what the colonists called the "Intolerable Acts," closing the harbor until the EIC was paid. The colonists disobeyed this order and began the coordinated resistance that escalated into the Revolutionary War in 1775. Like cotton and the Civil War, a single commodity – tea, was the flash point of armed resistance. It is perplexing that the British Parliament would allow such a war to commence, in large part to protect the power and prestige of a single company.

First Opium War

The East India Company had a major problem – the Chinese were really not interested in anything in trade except silver (Allingham 2006; Martin 2007; Mair and Hoh 2009). In the early 1700s, Europeans were able to get silver by trading sugar to Mexico, Peru and Bolivia, but these sources pretty much dried up during the Seven Years' War and the American Revolution. The EIC's solution was to build a demand for opium in China, sell it for silver and then use that silver to buy tea.

Of course, the problem with this plan was that opium use in China was banned by the emperor and the British government certainly couldn't advocate the trade of an addictive substance. In response, the EIC was forced to ban the transport of opium in any of their ships. However, they developed a circuitous route around this problem by growing poppies in India, processing it into opium, and then selling the opium for silver to private merchants called "country agents." These country agents smuggled the opium into the Pearl River Delta of China, bribed the necessary officials and sold it for silver.

What started as a minor trade item grew by leaps and bounds to gigantic proportions and opium soon become the most profitable trading commodity in the world. In 1760 only about 200 chests (140 pounds) were exported to China but by 1770 the amount had risen to over 1,000. By 1800 this level was at 4,000 chests, and by 1830 had reached over 18,000 annually. In 1830, there were likely over ten million opium addicts in China, all smoking away their money in opium dens. The opium-for-tea exchange came to essentially rule the British economy – at its peak about 10% of the country's tax revenue came from tea (Rose 2009). Opium financed the administration and governance of India.

The emperors in China became increasingly concerned with the vast number of Chinese that had become addicted to opium, and its import was banned. However, the opium trade was largely ignored due to its profitability. The first real action was taken by the Qing emperor in 1838, when he made Lin Zexu governor general of the Hubei and Hunan provinces and ordered him to abolish the opium trade.

Lin almost immediately made 1,600 arrests and gave the traders in the Canton estuary three days to turn over their loads of opium. When they refused, he closed the harbor and held them captive, not allowing them to unload or take on provisions. Tempers flared when a group of British soldiers destroyed a temple and murdered a man who tried to stop them. The British chief superintendent of trade in China, Charles Elliot, told the traders to hand over their opium to Lin, promising that the British government would compensate them. They balked at first, but eventually turned over 20,000 chests of opium that were subsequently destroyed.

A fleet of warships was sent to China in the summer of 1840 to break the blockade and it easily overwhelmed a Chinese Junk squadron outside of Hong Kong. The British fleet then moved up the coast to Chusan (Zhoushan), which they also captured easily after a nine-minute bombardment. To stop

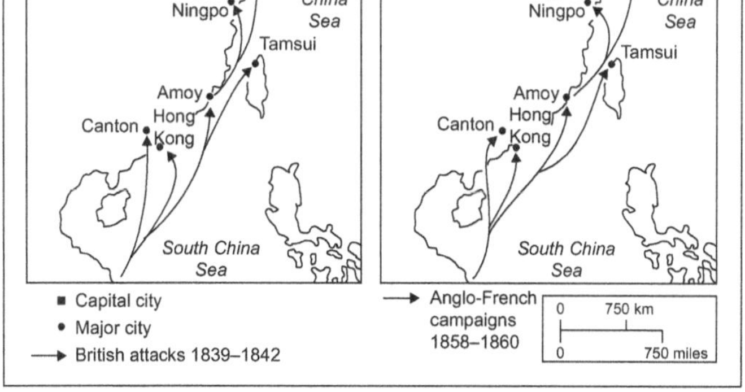

Figure 4.2 Maps of Opium Wars

Source: http://www.lasalle.edu.

the hostilities, the new imperial commissioner of Canton, Oishan, ceded Hong Kong to the British in return for Chusan. Both the British and Chinese governments were extremely dismayed by these actions and Oishan was put in chains and imprisoned for his surrender of Chinese land. Elliot was also relieved of command for promising compensation by the British government for the destroyed opium.

British Prime Minister Palmerston appointed Henry Pottinger to replace Elliot and force the Chinese to compensate the traders and open China to unrestricted trade. He arrived with a fleet of twenty-five ships, fourteen steamers, nine other support vessels and 10,000 men. He easily retook Chusan on October 1 and moved to Ningbo on the Chinese Mainland on October 13. There his soldiers attacked a force of 12,000 regular troops and 33,000 militias and rapidly dispatched them in a bloody slaughter. The weapons of the Chinese were no match for the firepower of the British; the front line of the Chinese consisted of 700 Sichuan soldiers carrying nothing but long knives.

As the British moved further up the Yangtze River, the shocked and overwhelmed Chinese realized they had no possibility of stopping the British advance. They agreed to the Treaty of Nanking in August 1842, in which they gave Hong Kong to the British, paid them an indemnity of $21 million, and opened five ports to open trade – Canton, Amoy, Fuzhou, Ningbo and Shanghai. The thoroughly humiliated Chinese had been forced to open their doors to the foreign dogs and submit to their military superiority.

Second Opium War

The Treaty of Nanking really didn't satisfy the commercial interests of the European powers and the United States (Allingham 2006; Martin 2007; Hickman, n.d.). Britain, in particular, wanted an ambassador stationed in Beijing, all ports to be open to trade, a fully legal status for opium and no taxes on imported goods. It was only a matter of time until the joint powers would exert their will.

They were given their excuse to act, when in 1856 Chinese officials boarded the Hong Kong-based ship *Arrow*, took twelve Chinese crewmen and threw them in prison. The charge was piracy and smuggling. The British demanded their release, arguing that the *Arrow* was not a Chinese vessel and fell under the Treaty of Nanking. Europeans were also riled by two additional Chinese actions – the execution of the French missionary Auguste Chapdelaine, and a plot by Chinese bakers to poison foreigners in Hong Kong.

To redress these complaints, the British asked France, Russia and the United States to form an alliance. A joint force of British and French attacked forts on the Pearl River south of Canton in 1858. These forts surrendered without a fight and the forces headed north. They took Canton after a valiant but futile resistance by locals and then won back the forts outside of Tianjin. To stop the advance, Emperor Xianfeng negotiated the Treaties of

Tianjin, allowing the British, French, Americans and Russians to establish legations in Beijing, open another ten ports, allow foreigners to travel all across the country, and to be paid reparations. In a separate agreement, the Russians also acquired a long strip of Pacific coastline in northern China where they built a naval base.

As would be expected, these treaties were not popular among the ruling elite of China and Xianfeng soon reneged on them by sending forces to the Taku Forts, and not allowing Admiral Sir James Hope to land troops to escort the ambassadors to Beijing. The British responded by bombarding the Taku Forts, but this time met heavy resistance and had to withdraw with help from an American fleet. Lord Elgin and General Charles Cousin-Montauban then arrived to enter the fray with over 170 ships and 10,000 Anglo-French troops. They retook the Taku Forts on August 21 and began to march toward Peking. As they approached the outskirts of the city, Xianfeng panicked and requested peace talks, but these were soon aborted when a British envoy and his party were arrested and tortured. The Chinese made a desperate final move to defeat the British-Anglo army by hitting them with wave after wave of humanity near Zhangjiawan, but failed miserably. The war ended when the much smaller Anglo-French force routed 30,000 Chinese troops defending Peking. In a final bit of retaliation, Xianfeng's Summer Palaces in Beijing and Chengde were looted and burned.

Xianfeng abandoned the city and left it to his Prince Gong to negotiate for peace. The Chinese were forced to accept the Convention of Peking where they agreed to abide by the original Treaties of Tianjin, allow open trade with Tianjin, ceded the Hong Kong city of Kowloon, permitted religious freedom, eliminated all restrictions on the opium trade and allowed the export of indentured laborers to the United States to work on the transcontinental railways.

The humiliation of Xianfeng greatly undermined the authority of the Qing Dynasty. Xianfeng's hurried flight out of Peking, his defeat by a much smaller western army, and the destruction of the Summer Palaces was a clear demonstration of the Qing dynasty's weakness. Confidence in the emperor plummeted. As described by Philip Allingham (2006):

> What had begun as a conflict of interests between English desire for profits from the trade in silk, porcelain, and tea and the Confucian ideal of self-sufficiency and exclusion of corrupting influences resulted in the partitioning of China by the Western powers (including the ceding of Hong Kong to Great Britain), humiliating defeats on land and sea by technologically and logistically superior Western forces, and the traditional values of an entire culture undermined by Christian missionaries and rampant trading in Turkish and Indian opium.

China would be the "Celestial Empire" no more.

Assam tea

As time went on, the East India Company realized that it would be advantageous to establish their own tea plantations somewhere and back away from the volatile Chinese market (Standage 2005; Mair and Hoh 2009; Griffiths 2011). Because they ruled India, it was a natural choice. In 1788, they went as far as to ask the famous British botanist Sir Joseph Banks to recommend crops for India, and he came back with tea as his number one choice. However, they backed off for the time being, not quite ready to abandon their profitable Chinese enterprise.

When the British government ended its Chinese monopoly in 1833, the company began in earnest to search for alternate sources of tea. The head of the East India Company and governor general of India, Lord William Cavendish Bentinck, set up a "Tea Committee" to study the feasibility of growing tea in India. They consulted with the Dutch who had been trying to establish tea in Java, and sent their chair, James Gordon, to China to obtain seeds, tools and tea workers. They also began the search for the place in India where tea could best be grown.

Little did they know that the tea plant *Camelia sinensis* was native to the Assam region of India, between the northern parts of Myanmar and southwestern China. The local Singpho people were even making tea from the bush. A government botanist in Calcutta, Nathaniel Wallich, had initially ruled that a sample of plants from Assam were just a run-of-the-mill Camelia, but he changed his mind in 1834 when he became a member of the Tea Committee and received another sample. The committee sent an expedition to the region and confirmed that *C. sinensis* was indeed native.

The East India Company now had to learn to grow and process tea in India. The so-called experts continued to argue over where it should be grown in the country and whether plants should be imported from China. A British adventurer, Charles Bruce, came to the rescue, relying on his hard-earned knowledge of the Assam region. The leader of the Singpho tribe provided him with native tea plants, and he absorbed all the knowledge he could get from the Singpho and a few experienced Chinese workers. He sent the first small shipment of tea to London in 1838, which the local merchants deemed highly satisfactory.

In 1840, with the Opium Wars in high gear, the Assam Company was established to grow and process tea in India. Investors literally came out of the walls and many had to be turned away. The company initially lost huge sums of money on the mistaken belief that the importation of Chinese workers was all that needed to be done to become profitable. It was not until 1847, with Bruce at the helm, that the company began to turn a profit. Similar to the old sugar factories, he industrialized tea production, picking and processing the crop on site. Production methods became increasingly streamlined through mechanization until the cost of tea processing in India was about one-quarter that in China.

Tea plantations also cropped up in other locations across India in the mid-1800s. Cachar became a major center of tea production in the southern horn of the Assam. A Dr. Campbell of the Indian Medical Services established a tea estate in Darjeeling at the base of the Himalayan foothills. Another doctor named Christie established the first tea plantation in southern India in 1832, about fifty miles inland from Calicut. Tea plantations were also planted in 1864 in the Travancore Mountains on the western side of India's southern tip, across from Ceylon.

Robert Fortune

While tea from the native plants of India proved acceptable to the British palate, the more delicate, aromatic teas from China still held sway. To build a really thriving tea industry in India, someone would have to go to China and bring bushes back. The person who accomplished this feat was the famous botanical explorer Robert Fortune (Rose 2009).

Fortune entered China after the Treaty of Nanking and traveled deep into the country where the best teas were grown. Animosity toward the British was high, forcing him to disguise himself by wearing native dress and shaving his head except for a dangling pony tail. He was almost totally dependent upon on his two Chinese servants for even the simplest needs and came close to losing his life on numerous occasions through disease, ambush and shipwrecks.

Being the curator of the Chelsea Physic Garden, he was interested in collecting all forms of plant life unique to the West, but tea and its manufacturing became the focus of his attention. His challenge was great, as the methods of production of tea were a closely guarded secret and the export of tea plants was strictly prohibited. In spite of these obstacles, he learned all things about tea. On his first trip in 1843 as an agent of the Royal Horticultural Society, he saw how tea was manufactured and he was the first Westerner to learn that green and black tea came from the same plant. On his second trip in 1848, he managed to collect and send 20,000 tea plants and seeds to India in two shipments for the East India Company. His first shipment met with disaster but a second one proved successful.

In his first attempt, Fortune sent 13,000 young plants in Wardian cases, glass boxes named after their inventor. The Wardian cases were designed to allow the plants to receive ample light, but kept their environment moist and optimal for growth. The bulk of the plants were in good condition when they arrived at their first Indian destination at the Calcutta Botanic Garden, but during shipment by river and oxcart to the company's experimental plantings at Saharanpur in the Himalayas, most of the boxes were damaged and only eighty plants survived. Undaunted, Fortune tried again by placing seeds rather than adult plants in the soil of the Wardian boxes. His hope was that the seeds would germinate on transit and the boxes would be filled with young plants on arrival. This approach worked and thousands of healthy plants found their way to the plantation!

The tea industry in India expanded dramatically from this point and the dominance of the Chinese industry soon ended (Martin 2007). Over two million tons of tea were shipped to England in 1862 and this number had risen to six million by 1866. Imports from India first surpassed China in 1889. As Indian tea production boomed, the Chinese industry was ravaged, contributing greatly to the economic and social chaos that befell China in the latter half of the eighteenth century.

Dutch East Indies

In 1827 the Dutch Tea merchant, J.I.L.L. Jacobson arrived in Java with the intent of growing and producing tea (Mair and Hoh 2009; Griffiths 2011). He started by making green and black tea from bushes growing at the Buitenzorg Botanical Garden, about forty miles south of today's Jakarta. While the tea was excellent, the bushes had come from chilly Japan and were not well adapted to the tropical climate of Java. Jacobson decided he would have to go to China to see if he could smuggle out better adapted bushes.

Over a period of six years, Jacobson made a number of trips to China to learn about their methods of tea manufacturing and to bring back bushes and experienced tea workers. He was essentially a tea spy, operating similar to Fortune but almost two decades earlier. On his fifth trip, he brought back a miraculous 300,000 bushes and twelve workers, ten of whom were subsequently murdered by rioting Javanese workers. In his seventh and last trip, now with a price on his head, he brought back an even more astonishing seven million seeds, fifteen workers and a cache of tea processing equipment. This time, he barely made it out of the country alive, as the authorities tried to chase down his vessel, capture him and seize his cargo.

For the next fifteen years, Jacobson settled down to cultivate his tea, shipped it to Europe and wrote the first definitive European treatise on tea entitled "Handbook for the Cultivation and Manufacture of Tea." When he died in 1848, the Java tea industry lost its heart and soul and stagnated, recovering its moxie in the late 1870s when the first Assam tea plants, with strong tropical adaptations, were imported into the country. By 1900, Java became the fourth largest tea exporting country in the world after India, China and Japan.

Tea in Ceylon

Tea production was in full force in India by the mid-1800s, but it took a couple of decades longer for tea plantations to take hold in nearby Ceylon (Sri Lanka) (Fernando 2000; Mair and Hoh 2009). It wasn't until the coffee rust virus started to decimate this acreage (see Chapter 6) that distressed plantation owners began to turn their eyes toward tea. By the early 1880s almost all the coffee bushes had been wiped out and a replacement crop was desperately needed. The growers first experimented with the quinine-producing cinchona tree with good results, but ultimately they turned to tea. Overproduction reduced the value of quinine to the point that the cinchona tree was no longer profitable to grow.

The first tea plantation, the Ceylon Company Limited, was begun by James Taylor in 1867 on nineteen acres in the central highlands of Ceylon, eighteen miles south of Kandy. His Loolecondera Tea Estate was planted with Assam bushes he obtained from the Royal Botanical Garden in Peradeniya. Tea estates gradually popped up all across the highlands, with fetching names such as Bobahawatte, Lover's Leap, Pooprassie and Thotulagalla. All the estates were quite isolated and the processed tea had to be moved by bullock carts down treacherous, winding roads with multiple switchbacks to Colombo, where it was sold and shipped to Britain.

The movement toward tea in Ceylon started slowly and then exploded. There were only about 350 acres in 1874 and 1,080 in 1875, but by 1883 the acreage had risen to 32,000 acres and in 1895 the total was 305,000. The first Ceylon tea to find its way to London was from the Loolecondera Tea Estate in 1873, totaling about twenty-three pounds. Only about 1,000 pounds of tea was exported in 1878, but by 1884 Ceylon was shipping 2.5 million pounds of tea, which rose to fifteen million pounds in 1887 and twenty-eight million pounds in 1889.

Trials of establishing tea gardens

The first tea planters in northern India faced almost unbelievable hardships. The tea gardens were extremely remote – Nazira, the headquarters of the Assam Tea Company, was 1,000 miles from Calcutta (Wickramasinghe and Cameron 2005; Mair and Hoh 2009; Griffiths 2011). To get to their new plantations sites, the growers had to bushwhack across dense forests after making long lonely trips down the mighty Brahmaputra River. Most planters got malaria immediately upon arrival in India, and they were further menaced by dysentery, cholera, typhoid and countless parasites. They faced extreme heat and humidity, torrential rains, hurricanes and even earthquakes. The land they had to clear was well populated with poisonous snakes and animal predators of all kinds including leopards and tigers. The planters and their families were also subjected to periodic raids by savage local tribes, who murdered the men and carried off the women. The only respite these pioneers had was to occasionally gather with their brethren at distant locations for boozing and socialization.

Another huge obstacle they faced was finding workers to clear the land, plant the bushes, pluck the leaves and process the tea. The locals were small-scale subsistence farmers who had little interest in working tedious, long hours for foreigners. The planters of Assam first tried importing thousands of Chinese tea workers, but this experiment proved to be a disaster. Many strayed or died on the way to the tea gardens, and those who did arrive proved to be unruly and inexperienced with picking Assam tea bushes and often got into bloody squabbles with the local villagers. The ultimate solution to the labor problem was to import workers from Bengal and Bihar, several hundred miles distant. Workers were also brought from closer Nepal and Nagaland, but the bulk of the workers came from India.

Hundreds of young elephants were also forced to work on the tea plantations (Martin 2007). Their brute force was critical in clearing the forest and transporting the tea through the dense forests of India and Ceylon. Their natural numbers dropped alarmingly through capture and habitat loss, and today they are ten times closer to extinction than the African elephant.

Recruitment of plantation laborers

The workers were recruited by contractors or "sirdars" who also worked in the tea gardens. Plantation workers would be sent back to their distant home villages with a little money to entice prospective recruits. They convinced family, old friends and neighbors to sign on as indentured servants (coolies), often providing cash advances on interest. The contractors got paid by the head, whether the workers arrived at the plantations alive or dead. These ruthless recruiters focused on numbers rather than quality, and anywhere from 10% to 50% of their recruits died en route.

Upon arrival, the coolies were organized into work gangs under a "kangany." These could be the sirdars themselves, or influential people of higher castes. The kanganies served as intermediaries between the plantation management and workers. They were paid a daily bonus for each worker that came to work, and they were often the paymasters. They became money lenders, shopkeepers and distributed rice to the workers for their sustenance. All in all, they became a key component of the north Indian plantation system, much more powerful than the overseers used for cotton and tobacco in the United States who ruled only by intimidation and physical force.

Figure 4.3 Coolie tea pickers in Ceylon

Source: http://www.wikimedia.org.

The recruitment of workers was also a major issue in Ceylon and the Dutch East Indies. It was less of a problem in Darjeeling at the foot of the Himalayans where workers were available from nearby Nepal. The workforce in the tea gardens of Ceylon became the Tamils of Southern India who were first employed and were separated by only a few miles of ocean. By 1847, there were 47,000 Tamils working in Ceylon and by the turn of the century this number approached 300,000. The labor pool in the Dutch East Indies was recruited largely from the British colonies, Malaya and Singapore.

Many of the Tamils who worked the tea gardens of Ceylon were not indentured and returned annually to southern India during rice planting season. They walked along what was called the "North Road" from Kandy to the coast, where they boarded small boats to cross the twenty-mile Palk Strait. The route was extremely arduous and was really little more than a path through almost impenetrable, disease-ridden jungle. Many of the coolies got cholera and collapsed along the route, being left to die on the roadside.

Hard life of the coolies

The recruitment system was rigged so that the coolies could never escape the tea plantation (Wickramasinghe and Cameron 2005; Mair and Hoh 2009; Griffiths 2011). It was a rare laborer who did not carry some level of debt that they would never be able to repay. Most of them had gotten a cash advance at employment and after arrival they soon fell further in debt to the kanganies for various necessities of life. An interest rate of 120% per annum was not unusual. They were charged for days off due to illness or dalliance. Often, a loan was made just before the coolie's contract expired keeping him on the farm until it was paid off. Another trick was to make husbands and wives contracts non-concurrent, so that one would wait and stay on the farm until the other was released.

An appalling number of coolies died on the plantations and never even had a chance to leave. In a Royal Commission study of 1906, it was found that over a three-year period in the 1860s, of 85,000 coolies that arrived in Assam and Cachar, 30,000 died. Another survey in 1868, found an overall mortality rate of 26% in Indian plantations, ranging from 14% to 56%. In Ceylon, the mortality rate of tea workers was a little better, but in 1891 it was three times higher than the population as a whole.

The number one killer of coolies was disease, in particular malaria and cholera, aggravated by contaminated water, poor food, home sites built on swampy land, insufficient medical attention, malnutrition and overwork. The British government in India passed many laws requiring decent living and working conditions for the coolies, but they were routinely ignored. The isolated plantation owners wielded more power than the judicial system. The few complaints that reached the British authorities generally fell on deaf ears or resulted in light punishment.

It was also not uncommon for coolies to be physically abused. The most minor transgressions could lead to flogging, which was particularly violent for those who tried to leave the plantation before their contract had expired.

Public displays of humiliation were commonplace at many tea gardens. Mr. Samuel Smith of Flintshire described one such incident at the House of Lords in 1890:

> both men and women had been flogged; in the case of women, that they had been tied to a post in the porch of the manager's house, their clothes lifted up to their waists, and that they had been beaten on the bare buttocks with a stirrup leather by the orders of the assistant manager.
>
> (Hansard, House of Commons Debate, March 3, 1890, vol. 341, cc1632-6)

The terse response of the undersecretary of state for India was "such complaints have been made, and have been carefully investigated by the Chief Commissioner of Assam and reported by him to be greatly exaggerated."

The levels of abuse and mistreatment of the coolies varied widely from plantation to plantation, and in general the laborers were much better treated in Ceylon and Cachar than Assam and the Dutch East Indies. At the most humane run plantations, laborers were rarely flogged and were provided with free, clean housing, decent food rations, reasonable wages, free medical help, schools, nurseries, their own kitchen gardens and even temples. The plantations became the coolie's whole world and final resting place.

Many of the coolies did eventually leave the plantations to settle locally, but they continued to be harassed and intimidated. In particular, the resident populations of Tamils in Sri Lanka (formerly Ceylon) have long been persecuted, leading to a bloody civil war that was only recently resolved.

Shifts in worldwide tea production

Tea production began to drop precipitously in China in the 1880s, and by 1944 the country was no longer exporting any tea. India and Ceylon had become the major suppliers to Great Britain (Standage 2005; Mair and Hoh 2009).

East Africa became one of the major tea producers in the middle of the nineteenth century. The first tea plants found their way to Africa in 1850, and by the end of the nineteenth century tea was being commercially grown in Malawi, Uganda, Tanzania, Kenya, Mozambique and Zaire. Kenya eventually became by far the leading producer, first from large, state-run plantations and after independence in 1963, from predominantly smallholder plots. There are now over 400,000 small tea farms in Kenya, averaging an acre in size.

Tea production in China was still staggering when Mao died in 1976, but under the more moderate regime of Deng Xiaoping the industry was restored. The rebirth began on state-run plantations, but now is largely on smallholder farms. China produced 513 million pounds of tea in 1976 and this rose steadily to a billion pounds by 1986; China regained world leadership in 2005.

Tea innovations

The tea bag was first patented in London in 1896 by A. V. Smithe, although the first tea bags were sold in the United States in the early 1900s by Thomas Sullivan. He started with silk bags, which proved to be too dense to allow for proper infusion, and replaced them with cotton gauze, which worked much better. The Americans took to the convenience of tea bags quite readily, and by the middle of the nineteenth century, most US tea was brewed in bags. The more tradition-bound British took much longer to move from loose tea to bags. By 1968 only 3% of the tea in Great Britain was brewed in bags, and in 1971 the percentage was still only 12%. The Brits eventually gave way, and today about 90% of the tea in Great Britain is sold in bags.

Systematic tea breeding is a relatively recent development (Chen, Apostolides and Chen 2012). It wasn't until the late eighteenth century that vegetative propagation of tea was developed and new fields were planted with only elite types. Tea breeding did not begin until 1950 in Japan and Sri Lanka, and 1985 in China. Almost all the tea varieties now grown in Japan were bred and clonally propagated, but in China, India and Sri Lanka the percentage of improved varieties is much less, varying from 46% to 60%. Two of the leading varieties in China are still wild selections and have been grown for over a century – "Tieguanyin" (1780) and "Fuding Dabaicha" (1857).

During World War II, tea played an important role in the British war effort. British soldiers and sailors were offered tea at every possible opportunity, as a small respite of comfort and normalcy. Mobile tea centers rolled through the devastation of London during the Blitz, serving the brew to warm the hearts and heads of the resilient occupants.

Table 4.1 World tea production in 2010

Country	*Production (kilotons)*	*Percentage*
China	1,475.1	35
India	966.4	23
Kenya	399.0	10
Sri Lanka	331.4	8
Vietnam	157.0	4
Turkey	148.0	4
Indonesia	129.2	3
Japan	93.0	2
World total	4,162.3	

Source: Chen, Apostolides and Chen 2012.

Tea today

Of the tea produced in the world, 80% is black tea, 18% is green tea and 2% is oolong (Chen, Apostolides and Chen 2012). Black tea is most popular in India, Great Britain and the other European countries, while green tea is most popular in China, Japan and the Middle East. Oolong tea is drunk mainly in China. The top eight producing countries are China, India, Kenya, Sri Lanka, Vietnam, Turkey, Indonesia and Japan (Table 4.1). People in Kuwait drink the most tea per capita, followed by Qatar, Turkey, Afghanistan, the United Kingdom and Russia. Tea became popular in the Middle East due to the Qur'an's prohibition on intoxicating spirits; the largest producer in the region is Turkey.

The treatment of tea workers in India and southeast Asia still generates outcries. In a recent BBC story, "The Bitter Story behind the UK's National Drink," Justin Rowlatt and Jane Deith (2015) uncovered that

> in Assam, north-east India, found workers living in broken houses with terrible sanitation. Many families have no toilets and say they have no choice but to defecate amongst the tea bushes. Living and working conditions are so bad, and wages so low, those tea workers and their families are left malnourished and vulnerable to fatal illnesses. There was also a disregard for health and safety, with workers spraying chemicals without protection, and on some estates, child labor being used.

Similar charges have also been made by others about tea estates in Sri Lanka and Malaysia. In response to the BBC charges: "Several of Britain's biggest tea brands, including PG Tips, Tetleys and Twinings, have said they will work to improve the tea estates they buy from in India."

In the last few decades "fair trade" movements have emerged that certify tea produced in an environmentally and socially conscious way (Raynolds 2000; Martin 2007). For certification, the producers must be democratic associations of small growers or larger plantations where the workers are free from discrimination, have equal pay, can collectively bargain, have safe and healthy working conditions and there is no forced child labor. The producers must also protect forests and wildlife habitat as much as possible and minimize chemical fertilizer and synthetic pesticide use. The fair trade teas now hold a profitable niche of the market in many European countries and the United States, although the majority of tea consumed is not certified.

References

Allingham, P. (2006) *England and China: The opium wars, 1839–60*, Victorian Web, Lakehead University, Thunder Bay, Ontario, http://www.victorianweb.org/history/empire/opiumwars/opiumwars1.html.

Chen, L., Apostolides, Z. and Chen, Z-M. 2012. *Global tea breeding: Achievements, challenges and perspectives*. Springer, Heidelberg.

Fernando, M. (2000) *The story of Ceylon tea*, http://www.historyofceylontea.com/

Griffiths, J. (2011) *Tea: A history of the drink that saved the world*, André Deutsch, London.

Hickman, K. (Undated) *Second Opium war: Overview and history*, http://militaryhistory.about.com/od/battleswars1800s/p/secondopiumwar.htm.

Mair, V. H. and Hoh, E. (2009) *The true history of tea*, Thames & Hudson, London.

Martin, L. C. (2007) *Tea: The drink that changed the world*, Tuttle, Tokyo.

Mondal, T. K. (2014) "Introduction." Chap. 1 in *Breeding and biotechnology of tea and its wild species*. Springer, New Delhi.

Philbrick, N. (2013) *Bunker Hill: A city, a siege, a revolution*, Penguin Books, New York.

Raynolds, L. T. (2000) Re-embedding global agriculture: The international organic and fair trade movements. *Agriculture and Human Values*, vol. 17, 297–309.

Rose, S. (2009) *For all the tea in China: How England stole the world's favorite drink and changed history*, Penguin Books, New York.

Rowlatt, J. and Deith, J. (2015) The bitter story behind the UK's national drink, *BBC*, http://www.bbc.com/news/world-asia-india-34173532.

Standage, T. (2005) *A history of the world in six glasses*, Bloomsbury, New York.

Wickramasinghe, D. W.A. and Cameron, D. C. (2005) *British capital, Ceylonese land, Indian labour: The imperialism and colonialism of evolution of tea plantations in Sri Lanka*, https://www.researchgate.net/publication/242263010.

5 Tobacco

Beginnings

The archeological history of tobacco is sketchy at best, with few ancient records (Goodspeed 1954; Gerstel and Sisson 1995). The oldest evidence of tobacco use comes from a temple in the state of Chiapas, Mexico. Dated AD 432, there is a bas-relief that depicts a priest smoking a pipe during a religious ceremony. The second oldest evidence comes from AD 650 from a Pueblo Indian site located in Northern Arizona. There, loose tobacco was actually found with a pipe, and analysis showed it contained nicotine.

It is likely that tobacco was utilized by humans long before these records. Most archeologists believe that tobacco was first domesticated 5,000 to 7,000 years ago. Native species of tobacco are found all across the Americas that were likely utilized by humans in antiquity. The most commonly grown tobaccos, *Nicotiana tabacum* and *N. rustica*, originated in the Peruvian/Ecuadorian Andes, and had reached eastern North America before 2,500 BC.

How people came to domesticate *Nicotiana tabacum* and *N. rustica* is an intriguing puzzle. Both of these species are not found in the wild and are hybrids of two different progenitors found in the South American Andes. *N. rustica* is likely a hybrid of *N. paniculata* and *N. undulata*, while *N. tabacum* is a probably a hybrid of *N. sylvestris* and *N. tomentosiformis*. It is not known whether humans domesticated rare hybrids from the wild that are now extinct or they exploited hybrids that appeared in fields of their progenitor species.

Further adding mystery to the origin of these tobacco species is that they have four sets of chromosomes, while their progenitors have only two. The chromosome content of the early types of *Nicotiana tabacum* and *N. rustica* must have naturally doubled in a very rare evolutionary event. Adding even more mystery to the tobacco story is the existence in both progenitor species of *N. tabacum*, of an enzyme that neutralizes nicotine. Somehow, a rare mutation was incorporated into the early forms of *N. tabacum*, to provide humans with the compound they craved in tobacco.

Whatever the serendipitous origin of nicotine-laced *Nicotiana tabacum* and *N. rustica*, they were most likely quickly embraced by native peoples after their discovery and spread hither and yon. By the time of Columbus's arrival in the New World in 1492, they were distributed throughout South America, Central America, North America and the Caribbean Islands. *N. tabacum* and *N. rustica* had become a mainstay of all native American life.

Early uses of tobacco

No one knows for sure why humans began to use tobacco, but it is clear that it became a major part of life soon after its adoption (Gately 2001; Charlton 2004; Groark 2010). It was snuffed, smoked, swallowed, used as an enema and spread over the body. It played a central role in social and religious functions. Tobacco was used to cure a wide array of ailments including headaches, colds, catarrh, diarrhea and wounds. Smoke was puffed over soldiers' faces before warfare and blown over fields before planting. Its powder was mixed with lime or chalk and used as toothpaste. It was smoked or snuffed by common folk in low doses for a mild intoxication and in large quantities by shaman to put them in a stupor before religious activities.

Many different implements were used to administer tobacco by ancient peoples. Snuffing tubes are among the oldest artifacts associated with tobacco. Tobacco was wrapped in banana leaves or corn husks to fashion cigars for oral smoking. Pipes were used to inhale tobacco through the mouth and Y-shaped tubes were utilized to smoke tobacco through the nose. The Y-shaped tube was called a tabaco or tavaco, the name inaccurately given by Europeans to the plant itself. Proper native American words for *Nicotiana* depended on tribe and included betum, cogioba, cohobba, petum, picietl, quauhyetl or yietl.

The pipe became the tobacco delivery method of choice in North America. Thousands of pipes have been uncovered from remains of the Adena and Hopewell cultures, which spread tobacco from the Mississippi Valley to New York. Their first pipes were carved from a single stone and were bulky, probably only used in communal ceremonies; their later pipes were much more portable, likely personal and were intricately carved with birds, ducks, beavers and frogs.

The Mayans of today have maintained a rich heritage of tobacco use in Middle America. Tobacco is revered as an ally of people, working with them to ward off fatigue and pain, and protect them from outside forces. Snuff is the most esteemed form, carried in small, polished gourds and placed on the tongue or between the teeth and cheek for its pleasing, intoxicating effect. Ground tobacco in warm water is drunk to cure stomach aches, constipation and urinary blockages. Warmed poultices of wilted leaves are wrapped around body parts to heal broken bones, sprains and bruises. Powder is rubbed all over the body for protection while traveling at night.

The Europeans discover tobacco

When Columbus arrived in the New World in 1492, he and his crew became the first Europeans to be exposed to tobacco and the practice of smoking (Gately 2001; Charlton 2004; Burns 2006). Their first contact with the plant came when Columbus, suited in his best clothes, met a boatload of natives he assumed were Chinese. Understanding the importance of this meeting, the members of the local Arawak tribe offered Columbus and his entourage gifts of fruit, beads and dried tobacco leaves. Having no knowledge of tobacco, the Europeans accepted the beads and ate the fruit, but threw the leaves into the sea.

The Europeans first saw people smoke when a group was sent by Columbus on an expedition inland to find the supposed Chinese leaders and present them with royal letters of introduction. The party, of course, failed in this mission, but did meet up with a group of natives who were smoking. They had "a little lighted brand made from a kind of a plant whose aroma it was their custom to inhale" (as transcribed by Columbus chronicler Saint Bartolomé).

Two of Columbus's crew, Rodrigo de Jerez and Luis de Torres, actually tried smoking themselves, got hooked and became regular users. Rodrigo carried a supply of tobacco back with him on the return voyage and continued the habit, totally shocking those who witnessed him puffing. Poor addicted Rodrigo was charged and convicted by the Inquisitors for being possessed by the devil and was thrown into jail for three years.

As the Europeans continued to explore and conquer the New World, reports of tobacco use abounded. Gonzalo Fernández de Oviedo, the military governor of Hispaniola, described dismissingly how chiefs smoked tobacco through their noses for pleasure and how shamans smoked and drank tobacco to aid in their communication with the "devil." Before he put Montezuma to death, Cortés observed his custom of smoking for relaxation after dinner. The great explorer, Jacques Cartier, actually smoked a pipe with the Iroquois of Canada and described how its taste was peppery and hot. Amerigo Vespucci told of people placing snuff on their tongues in Brazil, which they had stored in a dried gourd.

Reports on the use of tobacco as a panacea also circulated widely in the 1500s. The explorer, Pedro Álvares Cabral, documented the use of tobacco in Brazil to treat a multitude of ailments including abscesses, fistulas, sores and polyps. The Spanish priest, Bernardino de Sahagún, described in Mexico how smelling fresh green leaves relieved headaches, rubbing powdered leaves inside the mouth cured colds, and crushed leaves healed neck lesions. Various other observers across the Americas noted that tobacco was being used to treat diarrhea, aches and pains, catarrh, burns and wounds.

In general, the first European reactions to tobacco smoking and snuffing was disgust. Tobacco use was considered smelly and likely injurious, and led to amoral intoxication and drunkenness. This attitude was not too surprising, as many Europeans, particularly the Spanish, were in the Americas to conquer or convert "savages" they felt were ignorant and inferior. Their tobacco use must have been aligned with the devil.

Early European use of tobacco

In spite of the European's early revulsion to tobacco, its potential medical use trumped religious based fears. It started in Spain and Portugal as a garden novelty, but by the late 1500s well-placed people all across Europe were extolling its health benefits (Billings 1875; Gately 2001; Burns 2006). The French ambassador to Portugal, Jean Nicot, was the first to widely proclaim tobacco's healing properties. An experimentalist at heart, he began his work by treating a tumor on a Lisbon man with an ointment prepared from tobacco. He then cured one of his pages for an ulcerated leg, a woman whose face was infected with ringworm, and a cook who almost cut his thumb off. He sent seeds and plants to the Queen of France, Catherine de' Medici, proclaiming its curative properties and recommending the use of snuff as a prophylactic. The French court rapidly endorsed its use, calling tobacco the "Nicotian Herb."

A doctor in Seville named Nicolás Monardes rocked the European world in 1565 when he produced the pamphlet "Joyful News of our Newe Founde Worlde." In it he outlined the many aliments that tobacco could cure including bad breath, difficulty breathing, toothache, kidney stones, tapeworm, dandruff, animal bites, maggot infections and essentially all medical problems, inside and out. Monardes's pamphlet was translated into English, French, Italian and Latin. The French version was cleverly dedicated to the Queen of France, with a frontispiece engraving of a tobacco plant titled "Herba Medici."

Within about a decade, snuffing began to spread outside of the French court and the shipping of tobacco from the Spanish and Portuguese colonies became a profitable business. The Italians, Germans and Swiss soon joined the tobacco craze, along with the English and Dutch. Tobacco reached the Middle East through the Genoese and Venetian merchants, and Africa, China and Asia via the Spanish and Portuguese traders. The Portuguese and Spanish also introduced tobacco to Africa, where it was quickly accepted by the Muslims and virtually every native peoples across the continent. By 1600, tobacco was a part of the culture of even the most remote tribes in Africa.

Across most of Europe, tobacco was adapted for mostly medical reasons and to ward off disease, but the English and Dutch adopted tobacco for simple pleasure. The great sea captains, Sir John Hawkins and Sir Francis Drake, noted how native Americans smoked tobacco in pipes for pleasure and brought tobacco and the practice back to England. The most trusted advisor of Elizabeth I, Sir Walter Raleigh, played a key role in the British adoption of smoking. He was the one who convinced her and the royal

Figure 5.1 Sir Walter Raleigh

Source: The Project Gutenberg e-book of *Tobacco; Its History, Varieties, Culture, Manufacture and Commerce*, by E.R. Billings.

court to take up the habit and further popularized tobacco use by handing out pipes and pouches of tobacco at the popular hangout, the Mermaid Tavern. Gallants of the court came to use elaborate pipes and smoking paraphernalia, and developed theatrical means of smoking. The cost of tobacco kept the masses from smoking it for a few decades, but as the seventeenth century dawned, tobacco was popular everywhere and had become widely available in apothecaries, alehouses and grocers.

Smoking became widely popular in England but was not without powerful opponents. King James I likened smoking to almost devil worship and widely condemned it. He wrote in a pamphlet titled *Counterblaste to Tobacco*:

> tobacco is a filthy weed, and the custom is loathsome to the eye, hateful to the nose, harmful to the brain, dangerous to the lungs and in the black stinking fume thereof nearest resembling the horrible Stygian smoke of the pit that is bottomless.

He heavily taxed tobacco, increasing its tariff by almost 4,000% during his reign. King James and Sir Walter Raleigh were obviously at great odds, which eventually led to Raleigh losing his head.

Virginia tobacco

Almost all of the tobacco that found its way to England in the mid to late 1500s came from the Spanish colonies, largely through piracy and plunder (Billings 1875; Gately 2001). Sir John Hawkins likely took the first tobacco to England in 1565 from Florida after a slave-trading, piratical tour of the Caribbean. Sir Francis Drake brought a whole ship load of tobacco to England in 1586 during the Anglo-Spanish War that he had plundered from stores in Santo Domingo, Cartagena and San Agustín.

The price received for tobacco in sixteenth century England was exorbitant and out of the reach of all but the very wealthy. In 1593, the price for a pound of tobacco was over £4, when a mug of ale went for a penny. The rich and royal were the only smokers; most of the tobacco that found its way to England wound up in apothecaries for medical use.

The English began to grow their own tobacco in the 1590s, but demand for the weed soon outstripped supply. It was the colonies in America that ultimately came to the rescue. The first two colonies in the 1590s were dismal failures, with only a handful of people surviving the winter, but the Virginia

Company in 1607 managed to stay afloat. Out of the 105 colonists who made it to Roanoke, fifty-three were still alive after a year and these numbers were reinforced with another 400 hardy souls in 1609. It was here that the English first grew tobacco outside their borders.

The man responsible was John Rolfe, who got seeds of *N. tobaccum* from Trinidad. While everyone else was focused on food crops or unreasonable requests to develop a silk industry, Rolfe honed his skills in the art of growing and curing of tobacco. He sent his first small shipment to England in 1613 and caused quite a stir. His tobacco had a much sweeter and more aromatic flavor than that coming from the Spanish colonies.

Rolfe and his Indian bride Pocahontas took his first commercial crop to England in 1616 to great acclaim. The public was fascinated by the native Pocahontas and was enamored with the taste and smell of his signature tobacco. By then Rolfe had branded it "Orinoco." Unfortunately, Pocahontas died that year in England, but Rolfe returned to Virginia in 1617 with the economic future of the colony set. He arrived just in time for the first Thanksgiving festival, a fitting celebration.

Now, everyone in the colony began planting tobacco on any spare piece of land they could find. The planters dangerously neglected the culture of the staple food crops, corn and wheat, for the more profitable tobacco. Through famine and Indian wars, Virginia shipments of tobacco to England rose steadily from year to year. What was 20,000 pounds in 1618 became 60,000 in 1622, even though the colony had suffered an Indian massacre of a third of its population. The promise of tobacco profits became so great that King James dissolved the London Company that had founded the colony and reorganized the trade as a royal monopoly. Tobacco exports rose to 500,000 pounds in 1627 and to an astonishing 1.5 million pounds in 1629.

Arrival of slavery

The factor most significantly affecting tobacco production in the Virginia colony was labor (Billings 1875; Gately 2001). In 1620 there were only 900 people in residence and in 1624 there were only 1,275, even though there had been 3,500 immigrants during the intervening years. The harsh reality of the New World, dealing illness, hunger and massacres, dictated a life expectancy of only about six months.

The solvency of the nascent tobacco industry in Virginia took a sharp positive turn in 1619 when a Dutch trader brought the first twenty African slaves to the colony. Many more soon followed as the Dutch were more than willing to trade slaves for tobacco that they could profitably sell in Europe. Slaves allowed the colonists to greatly expand the tobacco acreage and provided many with a disposable income that greatly enhanced their lives. Life expectancies greatly improved and wives were brought from England to establish families. The first shipment of women arrived in 1619, being sold at a cost of 120 pounds of tobacco each. By 1640, population numbers had risen to 18,000 white settlers in Virginia. Tobacco was now the principal export crop in North America and would remain so throughout the seventeenth and eighteenth centuries.

The southern colonists were now set on that dark path of dependence on slavery for their livelihood. The treatment of slaves in North America never reached the inhumane, sadistic levels practiced in the sugar colonies, but bondage remained bondage. They were required to do as their owners wished, had no personal freedoms, were kept in poverty and were regularly disciplined by overseers for even the most minor of perceived transgressions.

The slaves in the early tobacco states were also better off than their future brethren in the antebellum cotton states. While tobacco cultivation was extremely arduous, the tobacco slaves worked in task-oriented gangs rather than for a set number of hours. Once their tasks were done, they had "free time" to tend their gardens, socialize and fuss about their meager homes. They lived in compounds supporting small groups of five to twenty-five that often came from the same part of Africa. This afforded them with linguistic commonalities and a sense of community, and in some cases a family life. Unlike in the Caribbean where most slaves did not survive to reproduce, slave numbers in North America actually increased through reproduction. This resulted in the vast pool of slaves that fueled the cotton industry in the 1800s.

Through the labors of their slaves, many planters became wealthy and dilatant. As Billings (1875) quotes a man named Hugh Jones:

> The common planters lead easy lives without much labor, or any manly exercise, except horse-racing, nor diversion, except cock-fighting, in which some greatly delight. This easy way of living, and the heat of summer, makes some very lazy, who are said to be climate struck. They are such lovers of riding, that almost every ordinary person keeps a horse; and I have known some spend the morning ranging several miles in the woods. . . . No people can entertain their friends with better cheer and welcome; and stranger and traveler is here treated in the most free, plentiful, and hospitable manner.
>
> (p. 56)

Over a period of just a few decades the colonists went from starvation and threat of massacre, to a life of leisure maintained on the backs of a subjugated peoples.

Tobacco growing in the colonies

Seedbeds were prepared in January or February by clearing land and hoeing it smooth. Seeds were sown in March, raked to spread them out and covered with pine boughs (Billings 1875; Cotton 1998). After about a month, the emergent seedlings were thinned to about four inches apart to give them plenty of space to grow. When the plants were about finger-sized in length they were transplanted to one- to two-acre fields. New fields were often planted among the stumps of what had been a lush forest. The brushwood would be burned before planting and manure would be spread across the field.

Figure 5.2 Taking a hogshead of tobacco to market

Source: The Project Gutenberg e-book of *Tobacco; Its History, Varieties, Culture, Manufacture and Commerce*, by E.R. Billings.

The seedlings were planted in hills separated by three to four feet. The seedlings did not always survive and the hills often had to be replanted. When the plants had grown to about two feet tall, two or three outside leaves would be removed and the tops of stems were broken off of most plants to encourage the growth of thicker, larger leaves. More leaves might be removed depending on soil fertility and the quality of tobacco desired. A few plants were left unpruned to produce seed for the next season. Weeds were removed weekly with hoes and by hand. Every plant was regularly examined for hornworms, which were plucked off and killed. Suckers developed at the base of plants, which were removed to enhance vigor.

When the plants stood 6–9 feet tall in late August or early September, they were ready to be harvested. An experienced planter knew exactly when plants were at the proper maturity to maximize quality. This was judged by leaf color, texture and pliancy. Because plants matured at different times, the harvesters went through the fields multiple times. The leaves were cut just above ground level with a sharp knife, and were left to wilt in the field for a few hours. Most of the harvesting was done between 10:00 a.m. and 4:00 p.m.; the early morning hours were avoided so that the dew could evaporate from the plants. Before 1618 the leaves were just left in the field after cutting and covered with hay, but by 1620 the leaves were taken to a drying house where they were hung to dry.

As the leaves cured, their color changed from green to a light tan. This process took 4–6 weeks. When the planter deemed the time was right, the leaves were spread on the floor to "sweat" for a week or two. The tobacco was then sorted for quality; the best tobacco was still pliable but not too moist or moldy. The leaves were then tied in "hands" of 5–14 leaves, left on a platform to further sweat and then sorted for shipment. The leaves were twisted and spun into balls of as much as 100 pounds and surrounded with canvas or wood. For long distance transport, leaves were commonly pressed into "hogshead" barrels, which were loaded on ships for transport to England. These barrels were rolled to local markets.

Smokeless tobacco in America

In the 1800s, the oral use of tobacco reigned supreme in the United States (Christian and Glover 1987). A distended cheek filled with tobacco was a national characteristic. Cowboys, presidents, factory workers and baseball players alike, all took up the habit. To catch the expectorated juice frequently flying, spittoons became a fixture in high and low places, from saloons to seats of government.

Tobacco was taken orally in three ways – as moist snuff, loose-leaf chewing tobacco and blocks or plugs. The use of snuff was a widespread practice of poor southern women who wet short sticks with saliva, dipped them into their snuff box and then spread the moistened powder along their gums. Men in all stations chewed tobacco; the practice started with loose tobacco in bulky packages and progressed to compressed rectangles of tobacco or "plugs." These plugs were originally flavored by pounding them into augured holes in sweet sapped trees, but later were produced commercially by curing tobacco leaves directly in sweet syrups (often molasses based).

Chewing was a distinctly American use of tobacco, actually eliciting ridicule and disdain from the British. Charles Dickens in his *American Notes* (1842) came down particularly hard on the practice, offering revulsion at the careless chewing and spitting of vile yellow streams. Dickens wrote:

> In the courts of law, the judge has his spittoon, the crier his, the warder his, and the prisoner his; while the jurymen and spectators are provided for as so many men who, in the course of nature must desire to spit incessantly. In the hospitals, the students of medicine are requested by notices on the wall to eject their tobacco-juice into the boxes provided for that purpose, and not to discolor the stairs. In public buildings, visitors are implored through the same agency to squirt the essence of their quids or "plugs," as I have learned in this kind of sweetmeat, into the national spittoons, and not about the bases of the marble columns.

At smokeless tobacco's peak of popularity, there were over 9,000 different brands of plug tobacco, more than 3,500 brands of chewing tobacco and over 3,500 brands of snuff. The public was bombarded with advertisements in the form of multicolored, graphic handbills, billboards and trading cards. By the Civil War, smokeless tobacco was a $22 million industry that was based primarily in the slaveholding states of Virginia, North Carolina, Kentucky and Tennessee.

Robert Koch in 1882 first isolated the bacteria that caused tuberculosis. Contemporary scientists soon realized that dormant forms of the bacteria could survive in a dried state, which meant that spitting could release infectious bacteria into the air. This helped fuel a growing public resentment against chewing and spitting in public places, although it took the rising popularity of the more dangerous cigarette to significantly reduce the use of

Figure 5.3 James Moran & Co.'s Indian Girl Chewing Tobacco, ca. 1874
Source: US Library of Congress.

smokeless tobacco. Smokeless tobacco reigned supreme until about 1913, when a blitzkrieg of advertising by R. J. Reynolds put their Camel cigarettes at the forefront of tobacco use in the United States.

Cigarettes take over

No one knows for sure where the first cigarette was rolled, but Turkish and Russian troops were smoking paper-rolled cigarettes by 1832 (McCusker 1988; Anonymous 2015). During the Crimean War in 1856, British and French soldiers picked up the cigarette habit and took the practice back to their home countries. Cigarettes soon crossed the ocean, and became a luxury habit in the antebellum United States. However, chewing tobacco remained king there for decades, with cigar smoking and snuffing holding on to secondary importance.

Cigarette smoking got a major boost in the United States after the Civil War. Northern and Southern soldiers congregating around Durham, North Carolina, became enamored with cigarettes made with a unique blend of tobacco cured in the area. What became known as "brights" were canary yellow and produced by "flue curing," where bursts of heat were used in the drying process. These tobaccos had a much milder flavor than the air-cured ones and were less acrid during smoking. When the soldiers returned home, they flooded the Durham area with orders for this unique tobacco, making

Caswell County, North Carolina, a center of tobacco production. The Bull Durham Tobacco Factory came to rule the roost.

The overwhelming dominance of the Bull Durham Tobacco Factory left little room for competition, but a family of gamblers and innovators managed to build a niche. When Washington Duke returned after the war from a Union prison, he was forced to sell his 300-acre tobacco farm. He started all over with his four sons by renting a few acres of his old property, and producing enough tobacco to make a modest profit. He and his sons used these earnings to build a cigarette factory, which didn't exist in Durham at the time. They started with experienced cigarette rollers from Europe, but soon negotiated the rights to a mechanical rolling machine invented by James Bonsack in 1881. The Dukes hired the engineer William O'Brian to get the machines functioning properly and were rewarded in 1883 with a factory line that had an output of 100,000 cigarettes a day. To package their cigarettes, the Dukes then developed a cardboard box with a sliding lid.

The Duke's factory could initially produce far more cigarettes than it had sales, but a marketing campaign touting the cheapness of his cigarettes soon built demand. Buck Duke loved to say: "Tobacco is a poor man's luxury. Where else can he get so much enjoyment for his five or ten cents." The Duke Company went on to dominate sales of the other tobacco products as well. They fought a "tobacco war" which led to the merger of five companies into the corporate giant, the American Tobacco Company. This "tobacco trust" completely dominated tobacco sales for the next two decades until it was broken up by an antitrust ruling in 1911.

The Dukes brought cigarette smoking to the masses, rich and poor. No longer were people simply puffing on pipes or cigars and relishing the aroma, but were now deeply inhaling cigarette smoke into their lungs. They were addicted. Cigarette sales boomed during the 1890s and during World War I smoking became a national habit. Cigarettes became associated with bravery on the front and were included in the soldiers' C-rations. Filters were added in the 1930s, preventing soggy butts and encouraging more women to smoke. Cigarettes were now a part of American culture, and Duke has a famous university in North Carolina named after him.

Tobacco health and the industry response

Since James I of England, there was a steady stream of medical, literary and governmental authorities who railed about the dangers of smoking, but none had been persuasive enough to seriously diminish the practice (Christian and Glover 1987; McCusker 1988; Gately 2001). In the 1920s and 1930s, medical practitioners became alarmed with the growing numbers of lung cancer victims in the United States. They became convinced that this growing incidence of lung cancer was linked to the precipitous rise in cigarette smoking. Thus began an almost a 100-year battle with the tobacco industry to admit the potential harm associated with smoking.

Four major studies in the 1950s and 1960s concluded that cigarette smokers had higher death rates than the non-smoking public and the rates of death were positively correlated with the amount of cigarette use. These finding led the US Public Health Service (1957) and the Royal College of Physicians in London (1962) to conclude that smoking did indeed cause lung cancer. Another study commissioned in 1964 by the US surgeon general supported the link between smoking and cancer, and associated the practice with other chronic broncho-pulmonary and cardiovascular diseases.

The tobacco industry's response to these reports was a vigorous, well-publicized series of denials and a massive advertising campaign that depicted smokers as virile, attractive and healthy. Liggett & Myers (Chesterfield) actually claimed they had sponsored a scientific survey that proved its brand was harmless. Doctors were featured in ads along with movie stars. Famous celebrities like Arthur Godfrey were recruited to assert that smoking hadn't hurt them; TV was replete with actors enjoying smoking and commenting on the pleasure the habit gave them. An aura was allowed to perpetuate that filter-tipped cigarettes took the "bad stuff" out, even though the filters themselves now contained tobacco by-products.

As America continued to puff, the number of cancer deaths kept rising – 2,357 in 1933, 7,121 in 1940 and 29,000 in 1956. The tobacco companies decided to fog the issue and joined together to publish a "Frank Statement to Cigarette Smokers," which appeared in 448 newspapers and potentially reached over forty-five million people. In this dictum they asserted that there was no direct evidence that smoking caused cancer and announced that they would establish a Tobacco Industry Research Committee to further investigate the issue. Cancer victims began suing the tobacco companies in the 1960s, only to be repeatedly squashed by the aggressive courtroom practices of the tobacco hired lawyers.

As time went on, more and more studies added to the list of illnesses associated with smoking. A list provided by A.W. Musk and N.H. de Klerk (2003) includes – "heart disease, stroke, peripheral vascular disease, COPD, cancers of the lung, oral cavity, larynx, esophagus, stomach, kidney, bladder, pancreas, cervix, vulva, penis, anus, and possibly others." They list additional

> smoke related health problems such as impaired fitness, peptic ulcer disease, premature skin ageing with wrinkles, osteoporosis, impaired fertility (including impotence) in males and females, snoring, abortion and ectopic pregnancy, inflammatory bowel disease, cataracts and increased complications of surgery, diabetes and hypertension.

In the 1970s, nicotine was shown to cause addiction.

By the late 1950s, the US government began to realize that trying to educate people about the dangers of smoking was not going to be enough. The number of smokers continued to rise dramatically. A warning label needed to be added to every pack. The first attempt to legislate labeling failed in 1957 under a withering tobacco lobby, but in 1966 Congress passed the Cigarette Labeling and Advertising Act, which put the words "Cigarette smoking can be hazardous to your health" on every pack. In 1971, cigarette advertising was also banned on all US television. Finally, in 1972, the proportion of men who smoked in the United States fell below 50% for the first time since the 1920s. The proportion of women smokers also plateaued and began to decline.

In another salvo in 1990, the US Surgeon General released a report summarizing over 57,000 studies, and declared that "it is safe to say that smoking represents the most extensively documented cause of disease ever investigated in the history of biomedical research." This report gave states the green light to sue the tobacco companies to recoup some of their ever-rising medical costs. By now, big tobacco had grown fearful of the possible extent of future state and federal suits.

In 1994, the attorney generals of Mississippi and Minnesota sued the tobacco companies for the Medicaid costs associated with the treatment of tobacco-related illnesses. A settlement was subsequently worked out behind closed doors that was subject to the passage in Congress of a bill limiting the liability of the tobacco industry for past actions. This bill failed, but the attorney generals of forty-six states did not give up and hammered out another "Master Settlement Agreement," without a legislative requirement attached. In this agreement, the tobacco companies would pay the states $206 billion over a period of twenty-five years. The money could be used by the states at their discretion, and future payments would be made based on volumes of cigarettes sold. The tobacco companies were also required to dissolve their organizations promoting cigarette use and stop targeting advertising to young people. In the end, the tobacco companies received a severe monetary penalty and were forced to act with greater responsibility, but were left standing to conduct most of their deadly business as usual.

Further tobacco use has been actively discouraged across the world by a wide array of national laws that increase prices through taxation and set bans on advertising, use by minors and smoking in public places (Ng *et al.* 2014). Most notably, 177 countries ratified the Framework Convention on Tobacco Control (FCTC) in 2003 and adopted all its proposals. These coordinated efforts have significantly reduced the percentage of worldwide smokers, but tobacco consumption is still estimated to take nearly six million lives each year and reduce the life span of smokers by 6.9%. Most health officials agree that the heaviest users, primarily low income individuals in developing countries, will need to be targeted to make more serious inroads into reducing tobacco consumption.

Tobacco today

While strong global efforts have been conducted by organizations such as the Food and Agricultural Organization (FAO) and the World Health

Organization (WHO) to discourage smoking (in addition to national laws), tobacco consumption has still increased worldwide by an annual average of 2% due to population growth (Anonymous 2003, 2010; Ng *et al.* 2014). The increase has been centered in developing countries where tobacco consumption has risen annually by 3.1%. In developed countries, tobacco usage has dropped by 0.2% a year. Among the developed countries, the decline is most significant in North America, where rates of smoking have dropped annually at an average of 1.3%. The proportion of adults in the United States who smoke has steadily gone down over the last sixty years; it fell from 50% in the 1950s to 40% in the 1970s to about 30% in the 1980s; today the percentage hovers around 18%.

Tobacco is now grown in more than 100 countries, with the major producers being China, the United States, India, Brazil, Turkey, Zimbabwe, Malawi and the EU countries Italy, Bulgaria, Greece, Spain and Poland. These countries produce over 80% of the globe's tobacco. In the last quarter of the twentieth century, there was a major shift in tobacco production from developed to developing countries. Production in developed countries fell 36% between 1970 and 2000, while in the developing world, production increased twofold.

Cigarettes are by far the predominant tobacco product worldwide, representing 85% of the world market. China consumes 31% of the global cigarettes produced, followed by the EU (mostly Italy, Bulgaria, Greece, Spain and Poland, totaling 12%), the former area of the USSR (9%), the United States (7%) and Latin America (mostly Argentina and Brazil, totaling 4%). In Africa cigarette consumption is still comparatively low, but is increasing steadily year by year.

Tobacco was introduced into China from the New World in the sixteenth century, and its use evolved through the centuries from long-stemmed pipes, snuff, water pipes, hand-rolled cigars and finally machine-produced cigarettes (see Benedict 2011). The American tobacco tycoon James Duke proved prescient in 1881 when he predicted that China, with its then forty-six million customers, would eventually be a major market for cigarettes.

References

Anonymous (2003) *Projections of tobacco production, consumption and trade to the year 2010*, Food and Agriculture Organization of the United Nations, http://www.fao.org/docrep/006/Y4956E/Y4956E00.HTM.

Anonymous (2010) *Trends in tobacco smoking 2000–2025*, World Health Organization, http://www.who.int/tobacco/publications/surveillance/reportontrends tobaccosmoking/en/index4.html.

Anonymous (2015). *Duke homestead: Cultivation of a tobacco empire*, North Carolina historical sites – Duke, http://www.nchistoricsites.org/duke/empire.htm.

Benedict, C. (2011) *Golden-silk smoke: A history of tobacco in China, 1550–2010*, University of California Press, Berkeley.

Billings, E.R. (1875) *Tobacco; its history, varieties, culture, manufacture and commerce*, American, Hartford, CT.

Burns, E. (2006) *Smoke of the Gods: A social history of tobacco*, Temple University Press, Philadelphia.

Charlton, A. (2004) Medicinal uses of tobacco in history. *Journal of the Royal Society of Medicine*, vol. 97, 292–296.

Christian, A.G. and Glover, E.D. (1987) History of smokeless tobacco use in the US. *Health Education*, vol. 18, 6–13.

Cotton, L.P. (1998) *Tobacco: Colonial cultivation methods*, National Park Service – Historic Jamestown, http://www.nps.gov/jame/learn/historyculture/tobacco-colonial-cultivation-methods.htm.

Gately, I. (2001) *Tobacco: A cultural history of how an exotic plant seduced civilization*, Grove Press, New York.

Gerstel, D.U. and Sisson, V.A. (1995) Tobacco, *Nicotiana tabacum* (Solanaceae). In: *Evolution of crop plants* (pp. 458–463). Smartt, J. and Simmonds, N.W. (eds.). Longman, Scientific & Technical. Essex, Sussex, England.

Goodspeed, T.H. (1954) *The genus Nicotiana*, Chronica Botanica, Waltham, MA.

Groark, K.P. (2010) The angel in the gourd: ritual, therapeutic, and protective uses of tobacco (*Nicotiana tabacum*) among the Tzeltal and Tzotzil Maya of Chiapas, Mexico. *Journal of Ethnobiology*, vol. 30, 5–30.

McCusker, K. (1988) Landmarks of tobacco use in the United States. *Chest Journal*, vol. 93, 34S–36S.

Musk, A. W. and De Klerk, N. H. (2003) History of tobacco and health. *Respirology, 8*, 286–290.

Ng, M., Freeman, M.K., Fleming, T.D., Robinson, M., Dwyer-Lindgren, L., Thomson, B., Wollum, A., Sanman, E., Wulf, S., Lopez, A.D., Murray, C.J.L. and Gakidou, E. (2014) Smoking Prevalence and Cigarette Consumption in 187 Countries, 1980-2012. *Journal of the American Medical Association*, vol. 311, 183–192. doi:10.1001/jama.2013.284692.

6 Coffee

Beginnings

The Ethiopians were the first people to drink coffee, perhaps 3,000 years ago (Allen 1999; Anthony *et al.* 2002; Wild 2004). They made their coffee from the species *Coffea arabica*, which is endemic to the southwestern highlands of Ethiopia. The first coffee drink was likely made from leaves rather than beans. The caffeine is much lower in the leaves, but an infusion of the leaves would still give a bit of a buzz like tea. Even today, Ethiopians drink two beverages made from leaves: kati or kotea, made from roasted leaves, and amertassa, made from leaves allowed to dry in the sun for a few days.

No one really knows when the first coffee beans were used to make a beverage, but legends abound. The story most often told involves a shepherd named Kaldi who observes that his flock of goats became frisky – dancing and mounting each other – after consuming beans from a coffee tree. When the goats didn't die, Kaldi himself tried eating the beans, felt a tingle and before long was frolicking alongside his goats. He passed on the miraculous effects of coffee beans to a local imam, who learned to boil the beans in water to make a drink that induced wakefulness during long overnight ceremonies.

Another popular tale is that a fellow named Ali bin Omar discovered the stimulating properties of coffee beans when he was banished into the desert outside Mocha in Yemen due to a dalliance with a king's daughter. Being left to die, he discovered that eating the beans from a tree gave him sufficient energy to stagger back to town. His astonished condemners allowed him to remain alive, believing God had sent him as a messenger to teach them about the power and utility of coffee.

By the mid-fifteenth century, coffee drinking was widespread in Yemen. The real person most commonly attributed with the invention of coffee is Muhammad al-Dhabhani, a scholar and visionary in the Sufi order of Islam. An important ritual in Sufism is the zikr, where proponents throughout the night chant repetitively and whirl about until they reach a state of ecstasy and are filled with God. Like the fictional imam learned in the story about

Figure 6.1 Kaldi dancing with his goats
Source: William Ukers (1922) *All About Coffee.*

Kaldi, Al-Dhabhani discovered that a brew made of coffee beans kept the worshipers alert all night and in communion with God.

Once embraced by the Sufi, coffee spread rapidly throughout the Arab world, and by the turn of the fifteenth century, it was being drunk in Arabia, Persia, Egypt, Turkey and North Africa. In the mid-1500s there were over 600 high and low coffeehouses in Constantinople. The Arabs began cultivating the trees in the mountains of Yemen, using all the Arab agricultural innovations including terraces and irrigation ditches. What had started as a religious drink was now a common part of everyday life across a vast area.

The Ottoman Turks occupied Yemen in 1556 and developed a vast trading network in coffee. Beans were shipped from Mocha to Suez, where they were carried by camel to Alexandria and then traded to French and Venetian merchants. The Turks kept their monopoly on coffee a closely guarded secret; no viable beans were allowed to leave Mocha without first being boiled or roasted.

Coffee – intoxicating or just invigorating?

As coffee began its spread across the Arab world and beyond, it became a social drink of the common folk and not just religious zealots (Wild 2004; Standage 2005). People could buy a cup on the street or in the market, and coffeehouses became all the rage in cities large and small. The growing popularity of the drink forced the Muslim leaders to take an introspective look at the nature of the physical effect of coffee – was it intoxicating like wine or harmlessly stimulating? This was an important issue, as Mohammed had banned intoxicants like alcohol.

In Mecca in June 1511, a local leader named Kha'ir Beg brought an expert panel together to decide whether coffee was indeed intoxicating. Composed of doctors, religious leaders and learned lay people, the panel tasted the brew and came to the conclusion that it was indeed intoxicating. Kha'ir Beg immediately banned coffee in the city, seizing and destroying existing stocks and having anyone beaten who dared to sell or drink it. His ban proved ephemeral, however, as it was overruled within a few months by a sultan in Cairo who was himself a hard-core coffee drinker.

What was most worrisome to the Arab authorities about coffee, was not the drink itself but where it was commonly drunk. The coffeehouses had become an important gathering place where political issues were discussed, along with rumors, gossip, news and all things of intellectual curiosity. Games of dubious morality to Muslims were also played there such as chess and backgammon. There were many attempts by various leaders to close coffeehouses throughout the sixteenth century, but none stuck. The popularity of coffee was simply too strong to allow any arguments about its intoxicating properties to prevail. Even Pope Clement VIII approved coffee drinking for Christians in 1601, remarking that such a delicious drink should not be enjoyed just by Muslims.

Coffee finds its way to Europe

The first coffee probably arrived in Europe about 1615, when Venetian merchants began importing it from Mocha (Ukers 1922; Wild 2004; Pendergrast 2010). It was consumed by relatively few Europeans until mid-century, when the first coffeehouses began to appear in England, France and Holland. Americans started drinking coffee in Boston in the late 1680s, but mostly in taverns serving also beer and tea.

The coffeehouses of Europe became what were referred to as "Penny Universities" where clients came prepared to conduct business and debate the topics of the day. The coffeehouses were a respectable, non-alcoholic alternative to taverns. They became centers of political debate where the overthrow of Charles I of England was plotted and the return of the monarchy in Charles II was supported. Camille Desmoulins is said to have jumped onto a table outside a French coffeehouse to beseech a mob to overthrow the aristocracy. The American Revolution was fermented by patriots who fueled their discourses with coffee in the Boston Green Dragon Inn. The consumption of coffee in the United States even became patriotic, replacing tea from Britain.

Debate raged across Europe in the 1660s and 1670s about the positive and negative benefits of coffee drinking. Some doctors called it poison, while others touted it as a cure for a wide range of ailments from headache to gout. Pamphlets were published that admonished people for frittering away their time in coffeehouses and having little time to service their wives. Complaints circulated that coffee drinking was too closely tied to Islam and that the coffeehouses served as sites of intrigue and rebellion. Charles II even tried to suppress new coffeehouses in 1675, believing them to perpetrate false and scandalous reports. However, public outcry to his decree quickly made him back down.

The Ottomans inadvertently contributed to the establishment of coffeehouses in Europe, both politically and militaristically. In 1699, an ambassador named Solomon Aga was sent by the sultan to the court of Louis XIV to foster an alliance to keep the Habsburgs at bay. The mission did not go well, as both the ambassador and Louis XIV managed to offend each other, but Aga set Paris Society agog when he began serving coffee socially in an opulent, rented palace that he had remodeled in the Ottoman style. When he went home, one of his cortege, Pascal, began selling coffee in the market of Saint-Germain to the proletariat, pulling in his clientele with the smell of coffee brewing. Pascal went on to open the first coffee shop in Paris, designing it like Aga's castle in the Ottoman style. Soon similar coffee shops were opened by other immigrants all across the city.

A siege of Vienna in 1683 by 300,000 Turkish troops also had a lasting impact on European coffee drinking. A man named Franz Georg Kolschitzky became a hero when he went across enemy lines to garner support from Polish and French troops. The Turks were forced to make a hasty retreat and left mountains of provisions behind including 500 bags of coffee. Kolschitzky was only too happy to accept the coffee beans as a reward for his services, and establish the first coffeehouse in Vienna. Having spent time in Arabian coffeehouses, he knew how to prepare the brew, and made it more palatable to European tastes by adding sugar and cream.

The monopoly ends

The Ottoman Turks were able to keep their Yemeni monopoly on coffee for about a century, but in the 1600s their hold began to weaken and break, when a few plants were smuggled out of the country (Ukers 1922; Anthony *et al.* 2002; Wild 2004; Pendergrast 2010). In the early 1600s, a Muslim pilgrim named Baba Buden strapped seven coffee seeds to his belly, and spirited the first viable beans out of Mecca to Mysore in southern India. The Dutch took descendants of these seeds to Java in 1699, established vast plantations, and spread coffee cultivation to Bali, Celebes, Sumatra and Timer. Soon, Dutch grown coffee began to rival the amounts and quality of that from Mocha.

Coffee reached the East Indies through another tortuous, but romantic route. Dutch traders pirated a single tree from Aden in 1616, which wound up in the Amsterdam Botanical Garden. In 1714, the burgomaster of Amsterdam gave some of its progeny to King Louis XIV, who had them planted in the Jardin des Plantes in Paris. One of these plants was procured by the French Naval officer Gabriel Mathieu de Clieu through court intrigue, whose dream was to establish a coffee plantation on his estate in Martinique.

During his passage home in 1723, de Clieu had to fight hard to keep his plant alive; protecting it in a box from raging storms, an attack by Turkish pirates and an envious fellow passenger. He even had to share his meager rations of water with his plant, when the ship became bogged down for a month by calm seas. When de Clieu finally arrived home, his plant flourished in his garden and over the next fifty years, nearly eighteen million of its progeny were propagated and planted on his estate and all across Martinique.

Coffee was also introduced by European traders from Martinique into San Domingo in 1715, Jamaica in 1730 and Cuba in 1748. The coffee tree was particularly well adapted to the higher elevations of the Caribbean Islands, where sugarcane would not grow but a ready workforce of slaves was available for its cultivation. By 1788, San Domingo was producing almost half of the entire world's coffee. This dominance came crashing down when the rebellion led by Toussaint L'Ouverture overthrew French control of the Island and put power into the hands of the former slaves. Rightfully fearing for their lives, the white coffee growers fled to Cuba where slavery still existed. Coffee exports from San Domingo fell in half, and Dutch Java coffee regained worldwide dominance.

Figure 6.2 Gabriel Mathieu de Clieu on board with one of his precious coffee plants

Source: William Ukers (1922) *All About Coffee.*

Ancestors of de Clieu's plant were further introduced to French Guiana from the East Indies in the late 1720s and from there were bootlegged into Para, Brazil, by Francisco de Melo Palheta. He had been brought in as an impartial mediator in a border dispute between French and Dutch Guiana, and while negotiating, had slept with the French governor's wife. She rewarded him for his favors by hiding coffee beans in the center of a floral bouquet that was presented to him as he set off to leave.

French traders also successfully smuggled coffee from Yemen in 1715 to the Isle of Bourbon (now called Réunion). Starting with forty trees, twenty died on route, another eighteen died after planting, but the remaining two thrived and were quickly multiplied to establish a coffee industry there. By the late 1720s, the island was producing 100,000 pounds of coffee a year. Progeny of these two plants were introduced to Kenya and Tanzania in the 1720s by missionaries, and eventually found their way to Brazil in the mid-1800s.

So, by the eighteenth century, coffee was being grown in the tropics of five continents. All the planting stock had originated from plants captured in only three, very limited thefts from Yemen – seven seeds to India, one plant to Holland and two plants to Réunion. The line of plants that passed through Holland and India came to called "Typica," while those traveling through Réunion were called "Bourbon." The narrow genetic base of all the world's coffee meant that every plant was vulnerable to any biotic challenge that might emerge anywhere. A new disease that killed one plant could kill them all. As will be related, this is precisely what happened in the late 1800s in Ceylon and Java, and eventually across the rest of the world.

Java coffee

The first coffee beans from Java reached Amsterdam in 1712 (974 pounds), and twelve years later almost 1.4 million pounds of coffee were being shipped to Europe by the Dutch (Thurber 1881). The number rose to over fifteen million pounds per year in the early 1800s and topped at over 200 million pounds in the mid-1800s.

The coffee industry in Indonesia was run almost as a government monopoly, with the native peoples being forced to work the plantations. There were some private growers but they amounted to only one-sixth of the total coffee crop. The means by which the Dutch obtained laborers was abominable. Instead of recruiting workers themselves, they paid the local chiefs to order their subjects to grow the coffee and then sell the harvest at a ridiculously low price to government agents. Families were assigned about 650 trees to maintain and harvest. As F. B. Thurber described the situation in his book *Coffee: From Plantation to Cup* (1889):

> The Dutch first conquered the island by force of arms, thereby establishing a claim to ownership of all the land, and then kept the native chiefs friendly by giving them a larger income than they formerly enjoyed, but requiring the chiefs in turn to make their people work systematically in cultivating the soil, and to sell their entire product to the Dutch government at a price so low that it barely gave them a subsistence.

Thurber further reported how the natives were treated with total disdain by the Dutch masters and were routinely humiliated.

> One feature of plantation life in Java would jar upon the sensibilities of most persons educated in American ideas and in the principles which lie at the foundation of government in the United States, viz., the contempt and want of consideration with which the natives are treated. As the natives approach the planter bearing some written message or report from the superintendents, they uncover their heads, and, bowing to the

> ground, they creep upon their hands and knees in order to hand the message to their master.
>
> (pp. 67–68)

Ceylon coffee

Dutch traders also tried to grow coffee in the coastal lowlands of Ceylon in the early 1800s, but it did not perform well there. Coffee remained largely a peasant crop until the 1830s and 1840s, when the British began to actively clear-cut and plant in the Kandyan Highlands (Thurber 1881; Duncan 2002). Two key pieces of British legislation stimulated this planting: (1) a law in 1835 that equalized the duty on coffee from Ceylon and West India, and (2) the Crown Lands Encroachment Ordinance Number 12 of 1840 that allowed Britain to seize any lands that were not under continuous cultivation and had no deed or tax receipt. The latter legislation essentially stole the land of the local Kandyan people, who had depended on shifting agriculture for ages across the region.

The British began removing what was an almost impenetrable rainforest, continuing the assault through the 1880s until all the highlands of central Ceylon were completely denuded and planted with coffee. The environmental devastation was tremendous. The thin soil remaining after the clear-cutting was greatly subject to erosion, and high winds now whipped across the hillsides tearing at coffee plants and uprooting many. All the native predators went extinct, allowing rat populations to explode across the plantations. The British built wind breaks and fought long and hard to stem the erosion, but were not deterred from destroying the whole forest. When the job was complete, well over 1,000 coffee plantations were spread across almost 275 million acres of the island's hill country. Coffee exports approached 100 million pounds annually.

The British planters expected the local Sinhalese villagers to be their labor force, but outside of helping clear the forest, the Kandyans were not at all interested in plantation work. The planters considered their refusal as sign of their innate laziness, but in reality, the Kandyans didn't need the work as they had subsistence farms that could support their families. Working on the coffee plantations would interfere with the harvesting of their own crops and they scoffed at the pitiful wages offered them by the planters. To man their coffee plantations, the owners had to turn to the Tamil people of southern India. The Tamils were a willing labor force for a number of reasons: British imposed peace in the region had led to a spike in the number of landless laborers, the area was routinely hit with famine, and the main source of income in the region – cotton weaving – had been superseded by cheap British cloth.

As was described in Chapter 4 on tea, the planters relied on the kangany system of worker recruitment and management on their coffee plantations.

In this system, Tamil recruiters went to their homes to entice workers to return with them and work on the plantations. The kangany system worked well for the planters as they did not have to recruit the workers themselves and they only had to pay the workers for the five-month growing season. They also did not have to pay for worker transportation, and because the Tamils were separated from their traditional support groups, they were much easier to control than the local Sinhalese.

As in all the plantation cultures, the Tamil workers or "coolies" (as they were called) were treated inhumanely, almost as slaves. They arrived at the plantations after a 150-mile march, already exhausted and hungry. It was not unusual for 20% to 25% to die on the march or soon afterward. The owners and overseers saw them as little more than animals, believing them to be child-like, lazy, indolent and sickly. The masters were convinced that only force would wring labor out of them, and they routinely beat and flogged the workers. They lived in filthy conditions, were given only enough food to keep them alive, and most of their medical needs were ignored.

The owners had decided that it was cheaper to recruit new Tamils to work the plantations than to take care of the ones already there. The Tamil laborers responded to this treatment by doing as little work as possible when masters' eyes were not on them, hiding when opportunity arose, damaging crops and jamming machinery to get a break. The forceful, callous management of the Tamils caused them to retaliate by becoming much poorer workers than they were naturally inclined to be.

The initial investors in Ceylon coffee were inexperienced government officials, clergy and business men. They made a tidy profit in the 1830s and early 1840s, but most went bankrupt in 1847, when the favorable duty they had received on coffee was removed and Europe fell into a depression. This group was replaced by new investors who better managed the plantations and operated at a profit even with lower prices. However, these owners too were wiped out a few decades later, when the fungal infection "coffee rust" or "left rust" began to devastate the fields. The causative agent *Hemileia vastatrix* marched relentlessly through the fields until all the coffee growers of Ceylon had to give up. The coffee was ultimately replaced by tea, as has previously been related.

Robusta to the rescue

Coffee rust soon found its way to Indonesia and decimated the Dutch plantations there. Rather than give up on coffee production as in Ceylon, the Indonesian growers found a resistant substitute species – *Coffea canephora*, or Robusta (Wrigley 1995; Pendergrast 2010). Native to central and sub-Saharan Africa, its beans were much smaller and more bitter than C. *arabica* but they were resistant to the coffee rust.

Robusta coffee had long been semi-cultivated in Africa, used as a popular shade tree and chewed as a "pick-me-up." It was first noted by explorers in Uganda in 1862 but only as a curiosity. They described how crushed beans were smeared with blood in Buganda tribal ceremonies. It wasn't until 1898, when it was rediscovered in the Belgian Congo, that its potential as a substitute for Arabica coffee was realized. It proved to be well adapted to the low, warm and humid lands of Indonesia, Ceylon and India where the Arabica coffee had failed due to rust.

Robusta coffee has a much coarser, harsher flavor than Arabica coffee, but it is a much more vigorous plant, is cheaper to produce and its bean contains almost twice as much caffeine. The plants of Robusta also come into full production in two years, as opposed to four years with Arabica. When it first appeared in the United States, it was banned by the New York Coffee Exchange because of flavor, but the Dutch quickly learned to love it. By 1920, over 80% of the coffee grown in Java was Robusta and today it is grown on over 30% of the global coffee acreage. It is most commonly used to make instant coffee, where its bitterness is not so apparent.

Slavery and the rise of the Brazilian coffee industry

Coffee became a major crop in Brazil at about the same time Emperor Pedro I declared their independence from Portugal in 1822 (Adams 1991; Wild 2004; Pendergrast 2010). By this time, the great sugar plantations in northeast Bahia had become non-profitable due to exhausted soil and plummeting productivity. In search of virgin land, the sugar growers streamed to the countryside north of Rio de Janeiro and São Paulo where millennia-old forests still remained with their rich humus. Many of these lands were too elevated and cool for sugarcane cultivation, but were perfect for coffee.

By 1850, coffee had almost displaced sugar in the São Paulo region and four times more slaves were toiling on coffee than sugar. Within a decade, sugar production was completely abandoned in the region. The economy of Brazil became centered on coffee, with about 75% of the country's wealth based on it. Half of the world's coffee was produced in Brazil in 1850, and that proportion rose to about 75% of the world coffee market in the 1900s. At least half of all the coffee produced in Brazil came from the São Paulo Province.

Vast tracts of forest were cleared to make great plantations or "faziendos" that contained two, three or even four million coffee trees and covered many square miles. As far as the eye could see, there was nothing but coffee trees. Similar to the sugar industry, the clearing of land and tending of coffee plantations was done on the backs of slaves. The white, privileged growers became wealthy treating their slaves just as inhumanely as the sugar growers. As Anthony Wild (2004) describes

Figure 6.3 Coffee slaves in Brazil

Source: Instituto Moreira Salles.

the situation: "Slave owning plantation owners were often fabulously rich, and inclined to inflict horrific suffering on their slaves: unremitting labor, near starvation, cruel punishments, murder and unreported deaths, rape, torture, sexual orgies. . . . Children, too, worked on plantation" (pp. 174–175). As with the sugar slaves, coffee workers lived an average of only seven years after they arrived in Brazil. Conventional wisdom among the sugar and coffee growers was that importing slaves was more economical than treating their slaves well enough to produce a future workforce.

The profitability of slavery led Brazil to maintain this sad institution for far longer than most other civilized countries. As previously related, a treaty banning the importation of slaves was signed with Britain in 1831, but was never enforced: over 200,000 slaves were brought into the country in the late 1840s. It took the presence of British warships in Brazilian ports in 1849 to finally entice the government to enforce the ban. Sadly, this action still left almost a million resident slaves to serve the plantation system for decades. More slaves were dying each year than being born, but the differential was only about 1% a year. At that rate, there would have been 30,000 slaves still left in Brazil in 1950.

As slavery became more concentrated in the coffee plantations, public pressure began to rise to free the resident slaves. In 1871, Emperor Pedro's prime minister, the Viscount of Rio Branco, got parliament to pass the "Law of the Free Womb," which freed the children of the resident slaves. This was another important step toward emancipation, but the still bonded slaves became increasingly restive. Thousands of them began fleeing the plantations in numbers that could not be controlled by troops and police. Plantation owners themselves began to see the long-term futility of slavery and started to emancipate their slaves. Finally, in 1888, the Princess Regent Isabel, acting while the aged Pedro II was out of the country, abolished slavery in the "Golden Law." Brazil's emancipation of the slaves followed the European countries by over fifty years and the United States by almost twenty years.

As it became clear that slavery would end, the Brazilian plantation owners began to entice poor Europeans (mostly Italians) to come and work the plantations as "colonos" or sharecroppers. They were given a home, a little land to grow their own crops, and assigned a number of coffee trees to tend, harvest and process. The colonos were, in fact, indentured servants who were required to pay off the cost of their transportation and any cash advances before they could leave the plantation. Most plantations had armed guards who kept the sharecroppers in place and in line.

The immigrants initially came in a trickle of thousands from 1850 to 1870, but between 1884 and 1914, over a million arrived. The Brazilian government greatly encouraged this migration by starting to cover the costs of their transportation in 1884. The former slaves preferred subsistence farming and occasional employment to working on the plantations, and the owners were more "comfortable" with the white workers they considered genetically superior to the former slaves. Under the colono system, coffee production boomed in Brazil, going from 5.5 million bags in 1890 to 16.3 million in 1901.

Coffee farming in Brazil

Coffee farming, similar to all plantation crops in Brazil (past and present), was slash and burn agriculture. Virgin forests were destroyed to plant the coffee, and when the soils were depleted of nutrients, the growers moved their plantations, leaving behind environmental devastation (Wild 2004). It takes a rainforest a century to reforest.

During the winter each year, slaves were sent out to girdle all the trees on a hillside. A few weeks later when the trees were dying, they were chopped down starting with the biggest ones in hopes the monsters would bring down smaller ones as they fell. The brush, trunks and stumps were then burned, covering the whole area for miles with a yellow, acidic smoke. The end product was a vast, denuded landscape covered with a nutritious ash.

The coffee trees were then planted in rows in the full sun, often up and down rather than across hills, encouraging massive erosion when the rains came. The trees took four to five years to reach full maturity and were productive for fifteen to thirty years, depending on how long the soil maturity held out. Harvest sizes were highly irregular from year to year as the trees had "mast" years of high productivity followed by low harvests the next year, as the tree essentially rested. In general, the coffee plantations of Brazil were not the world's most productive. Coffee plants perform best at 3,000–6,000 feet, where temperatures average 70°F and rarely exceed 80°F, but most of the coffee plantations in Brazil were below 3,000 feet.

Coffee was harvested in Brazil by the traditional dry method, where ripe and green berries were stripped off the trees. The beans were separated from tree debris by tossing the stripped berries into the air over screens allowing the wind to carry away sticks and leaves. The ripe berries were spread out on clay tile patios to dry, being turned several times a day to prevent spoilage. At night, they were raked together and placed under tarps to protect them from dew. After several weeks, when the berries had turned black, the husks were tediously removed by pounding them with stakes. During the drying process, care had to be taken to keep the berries well separated and away from external contaminants to prevent them from developing strong off flavors.

Coffee and repression in Guatemala

Before there were "Banana Republics" there were "Coffee Kingdoms." These Central American states were built by the brutal oppression of the native Mayan people (Lovell 1988; Pendergrast 2010). Upon declaring independence in 1821 from Spain, the first leader of the region, Mariano Gálvez, decided to build the economy by confiscating the land of the Mayans and turning them into sharecroppers and debt peons. Their children were taken and given to ladino "protectors" who often made them indentured servants. To escape bondage, the Mayans began migrating to the *altiplano*, the high elevation peaks, where the land was not as desirable for commercial agriculture but could still support indigenous crops.

Conditions temporarily improved for the Mayans in 1836 when Rafael Carrera led a revolution and split Guatemala from the rest of Central America. He was part Indian, and while he grew fabulously rich through graft and manipulation, he was sympathetic to the peasants and became their hero. During his reign, the economy of Guatemala was largely based on indigo and cochineal, a brilliant red dye extracted from insects that fed on cactus. The dye industry became doomed in 1856 when synthetic dyes were invented and an economic substitute had to be found. Coffee rose to fill the bill.

In the 1860s, it had been discovered that coffee was well adapted to the Verapaz highlands of the Pacific coast of Guatemala – land occupied by Indians and previously considered worthless. President Vicente Cerna, who had succeeded Carrera in 1865, remained friendly to the Mayans, but came under increasingly heavy pressure to build a coffee economy. In 1871 he was overthrown by a wealthy coffee grower, General Justo Rufino Barrios, a man obsessed with maximizing coffee profits through the repression and exploitation of the Mayan people.

The Barrios government immediately declared that any land not being used to grow coffee, sugar or cacao was *tierra balda* (idle) and was therefore state property. This began a huge land grab from the Mayans, where hundreds of thousands of acres of their ancestral lands were appropriated, parceled off and sold. In 1873 alone, 200,000 acres in the western piedmont were divided into 550-acre lots and sold at rock-bottom prices to ladinos willing to plant coffee.

The Barrios government also passed legislation to encourage foreign companies to invest and build in Guatemala. As a result, many Germans flooded into the country and many established huge coffee plantations, or *fincas*. The economy of Guatemala became dominated by the 100 largest fincas, which represented only 3% to 4% of the coffee plantations, but over 50% of the production. The small, family owned fincas became dependent on the larger ones for processing, and often carried huge debt loads, paying processors from 14% to 25% of their profits.

As if taking their lands was not enough, the Mayans were coerced into working the coffee plantations. Thousands of them were forced to harvest coffee each year, providing hundreds of days of free labor each season. Even the Mayans who had moved to the altiplano were forced to migrate down to the coffee fields during the harvest season. Many died of influenza and cholera, and those that survived took the diseases back to their villages.

Many of the frustrated Indians took to guerrilla warfare and were fought tooth and nail by Barrios's troops, who hunted them down and murdered them, forcefully resettled families who helped the rebels and destroyed homes and crops in the rebel zones. The whole country of Guatemala became almost a penal colony, dominated by a huge standing army and local militias that kept the Mayans working and under control. Sadly, this bloody strategy met its aims and the coffee economy of Guatemala boomed. In 1873, coffee exports were about 149 million kilos, and this rose to 691 million by 1895. At the turn of the century, Guatemala was exporting 100 million kilos.

The rest of Central America and Mexico

Coffee economies were also built on the forced labor of indigenous people in Mexico, El Salvador, Nicaragua and Honduras (Pendergrast 2010). Only in Costa Rica were natives not the primary work force, as most Mayans had already been exterminated during the Spanish invasion. The coffee fincas were smaller in these countries than in Guatemala, but the Mayans were just as brutally subjugated. Their ancestral, communal lands were confiscated and they were ruthlessly forced to harvest the crop. Men, women and children were all brought into the workforce, with the women and children doing most of the sorting and much of the harvest. Even the tiniest children trailed their parents in the fields, picking coffee from the lowest branches. Work conditions were most difficult for the women, who were often sexually abused and received far lower wages than men, if they got paid at all – they were often hired out by their husbands for cash advances.

Similar to Guatemala, most of the countries in Central America became bloody battlegrounds, with the oppressed Mayans repeatedly revolting, fleeing plantations and burning processing plants and fields. Military police and plantation militia were organized to fight the uprisings. In one particularly ugly incident in Nicaragua, thousands of Indians attacked government headquarters in Matagalpa in 1881, only to be brutally repulsed, ending with thousands of Indians dead.

The governments of Central American countries were constantly in a state of flux, subject to a revolving door of military dictators who overthrew one after the other. The situation was almost untenable for everyone, both ladinos and Indians alike, but no one knew an alternative to keep the coffee industry booming. As Mark Pendergrast (2010) put it:

> The coffee economy . . . frustrated everyone in one way or another. Above all, however, it relied on forced labor and misery of the indigenous population. With this unhappy foundation, a future of inequality and violence was all but assured.
>
> (p. 32)

The turbulent, coffee-fueled politics of Central America were also aptly described by Anthony Wild (2004):

> Coffee sustains their economies, but also sustains the entrenched land-owning oligarchies that control the industry, and hence the political and military machines that preserve the status quo . . . many of these coffee countries had been effectively slave states. The wealthy coffee plantation owners would, in cahoots with the government, coerce migrant labour down from the neighboring highlands. . . . Frequently the land that the coffee was grown on – by definition, fertile and at high elevation – was appropriated from indigenes who then became the first guerrillas to fight against the same government policies that had allowed their communal lands to be stolen. . . . The creation of wealthy land-owning coffee farmers, to whom the government was deeply beholden, created a cycle of suppression of political and the press, the use of the police and bureaucracy as an instrument of fear, and the replacement of any government that failed to behave in an accommodating fashion towards the coffee oligarchy. Under these circumstances, the only expression of opposition was out-and-out rebellion.
>
> (pp. 234–235)

Americans learn to love coffee

It is not known for sure when the first coffee arrived in North America, but it was likely consumed by the Dutch when they occupied New Amsterdam (New York) in the mid-1600s (Ukers 1922; Pendergrast 2010). The first written reference to coffee in the colonies was made in 1668, and it is known that William Penn purchased coffee for his Delaware settlement in 1683. Throughout the seventeenth century, most coffee was consumed in taverns and inns, where alcohol was much more popular. Dedicated "coffee-houses" did not begin to proliferate until the early eighteenth century, being most common in the larger cities of Boston, Philadelphia and New York.

Coffee was not popular enough during the Revolutionary War to be included in the rations of the Continental Army, but its popularity grew as an alternative to the hated British tea. Coffee consumption was further boosted during the War of 1812 when tea imports from England dried up and all things French became the rage. Per capita coffee use reached about 3 pounds in 1830, 5.5 pounds by 1850, and 8 pounds in 1859.

Most coffee in the first half of the 1800s was prepared over wood stoves in eastern homes or over the campfires of Americans migrating west. Green coffee was typically purchased at a general store, and was roasted in a frying pan. The coffee was then ground in a mill or with a mortar and pestle, and boiled in water for about ten minutes. Eggs and even fish by-products were sometimes added to help settle the grounds before drinking. A commercial roasting industry and two-tier drip pots did not emerge until the 1830s and 1840s.

It was during the Civil War (1861–1865) that coffee really became a centerpiece of American life. The blockade of southern ports kept it out of Confederate hands, but it was an integral part of the Union soldier's daily ration. Each soldier was given one-tenth of a pound of green beans a day, which would amount to about thirty-six pounds a year. Tiny fires spread out daily across Union campsites as the exhausted soldiers roasted and brewed this important pick-me-up.

Figure 6.4 Coffee call – Winslow Homer, 1863

Source: US Library of Congress.

Coffee prices actually grew out of the reach of the average American during the war, reaching the outrageous level of forty-two cents per pound, but at its conclusion, the price fell back to eighteen cents per pound. Worldwide production of the bean was now booming in Ceylon, Java, Brazil and Central America. Coffee prices would rise and fall over future decades in response to environmental disasters and price fixing, but most adult Americans were now sufficiently hooked to coffee's stimulatory effects that they were willing to pay almost any price.

The American coffee titan

John Arbuckle was the first man to build a coffee dynasty in the United States (Coppin 1990; Pendergrast 2010). He started as a grocery wholesaler with his brother and uncle in Pittsburgh in 1860, but in 1871 decided to focus on coffee. He and his brother built a coffee processing factory in New York, bought one of the newfangled coffee roasters of Jabez Burns and began selling coffee in paper bags all across the United States. To keep the coffee fresh and aromatic in transit, they infused it with a patented egg and sugar glaze. This likely didn't work but sounded good; such additives were popular at the time.

Under the primary leadership of John, the brothers' prepackaged, roasted coffee was an immediate hit. The production line in the factory started with fifty women manually putting coffee into bags but was very soon replaced by an automated packaging machine that did the work of 500 women. The brothers' brand of coffee, Ariosa, took the East Coast by storm and completely dominated the frontier states of the West. Ariosa became the coffee of choice of the cowboys. Even the crates used to ship the coffee became a part of Western life, serving as storage containers, cradles, coffins and even building materials.

John Arbuckle was a genius at marketing. He loved producing handbills that attacked his competitors and praised his product. In one he depicted bugs and filth in the barrels of a competitor, while in another, he showed a hassled housewife crying "Oh, I have burnt my coffee again." He offered a penny redemption for his signature on each package that could be used to purchase goods from a catalogue. The items available ranged from toothbrushes to clocks, guns and window dressings.

Arbuckle also worked hard to maximize profits by vertically integrating the company. He opened exporting offices in the 1880s in Mexico and the major ports of Brazil – Rio de Janeiro, Santos and Victoria. He built his own shipping fleet and bought a barrel factory that used wood from his own forests in North Carolina and Virginia. He set up warehouses in Kansas City and Chicago, and his flagship factory in New York grew to cover a dozen city blocks.

Arbuckle took no competitors lightly and if challenged fought them tooth and nail. In the late 1890s, he got into a pitched battle in what became known as the coffee-sugar war. For a long time, he had been buying all his sugar from H. O. Havemeyer's American Sugar Refining Company, but in 1897 decided to build his own sugar refinery, perhaps to lower his costs. Havemeyer took offense and went into the coffee business, slashed prices, and tried to bankrupt Arbunkle. Arbuckle responded by flooding the market with sugar to try to crush Havemeyer.

The coffee-sugar war lasted until 1903, when Havemeyer had to give up. He lost $15 million in the battle and was forced to sell his Lion Coffee company for only 40% of what he originally paid for it. Arbuckle was also hurt in the battle, but only to the tune of $1.25 million. Arbuckle won because of several competitive advantages. He sold sugar in small packages and directly to retailers, so obtained a higher price per pound. He also hired Joseph Stillman, a former employee of Havemeyer, to design a sugar refinery that could produce sugar at a much lower cost than Havemeyer's.

Arbuckle turned to philanthropy in the later years of his life and spent millions to find homes and healthy recreation for the homeless, handicapped and impoverished. The most inventive was to outfit ships with lodgings, recreational facilities, and dining rooms, and then take the poor out to sea to escape city life.

When Arbuckle died in 1912, this usually prudent man left no will and the company was probated to a nephew and two sisters. Speculation raged that one of them had actually destroyed a will that bequeathed all of his money to charity. Unfortunately, the family did not have Arbuckle's flair for business, and under their mismanagement, the company succumbed to competition and closed in 1935.

Coffee valorization in Brazil

By the mid-1800s, coffee was by far the chief export of Brazil, and production from there completely dominated the world market (Hutchinson 1909; Sensabaugh 1946; Pendergrast 2010). Over three-quarters of the global supply of coffee came from three states in Brazil – São Paulo, Rio de Janeiro and Minas Gerais. The availability of seemingly endless virgin, fertile soils and cheap labor allowed Brazil's coffee industry to thrive.

As the new century approached, world demand for coffee started to fall below production levels and surplus stocks became the norm. World production reached twenty million bags (132 pounds each) in 1901, while demand was only fifteen million bags. Warehoused stocks of coffee swelled to 11.5 million bags. Brazilian production rose from 9.5 million bags in 1899–1900, to 11.3 million in 1904–1905 and 20 million in 1905–1906. As a result of this overproduction, coffee prices began to drop precipitously. From 1880 until 1895, a pound of coffee in the New York market was worth from fourteen to sixteen cents per pound, but in 1901 the price had plummeted to six cents. World coffee stocks were so high that even a major frost in Brazil in 1901 did not affect prices over the next couple of years.

Something had to be done or all but the strongest planters would go bankrupt, taking the whole country with them. The government tried to prevent further planting by levying a tax of $180 on new coffee acreage, but this was too little, too late. An international coffee congress was convened in 1902, only to make toothless recommendations. In 1903, one of the largest coffee growers suggested that the only viable solution to the price crisis was for the government to join with private financers and merchants to buy up surplus coffee and hold it until prices improved. This scheme came to be known as "valorization" – holding the price of a commodity.

In 1906, the Brazilian coffee states signed an agreement to buy and hold stockpiles of coffee until prices improved. To do this, they had to borrow heavily from international banks, and Rio de Janeiro and Minas Gerais soon lost their nerve and dropped out. São Paulo, whose income was almost totally dependent on coffee, carried on and by the end of 1907 had borrowed the staggering sum of $88.4 million to valorize their coffee. Still, not enough coffee had been taken off the world market to significantly alter prices.

In desperation, the planters of São Paulo approached the German coffee merchant and entrepreneur Herman Sielcken and ask for help. He formed a syndicate of banks and coffee merchants, who agreed to take over most of the cost of purchasing and warehousing surplus coffee. The group would advance 80% of the sum needed to buy 500,000 bags a month at seven cents a pound. After a year and a half, this effort also had not substantially altered the price of coffee, and Sielcken was forced to form yet another syndicate. Finally, a corner was turned – the price of coffee rose to eleven cents a pound in 1910 and fourteen cents a pound in 1911.

At this point, the US government began to howl. A syndicate of coffee merchants was artificially manipulating prices and making the American people pay more for their cups of coffee! The first to scream bloody murder was Congressman George W. Norris of Nebraska who brought attention to the affair on the floor of Congress. The attorney general also got into the act, appointing William T. Chantland to investigate the nature of the valorization scheme.

The Department of Justice soon brought suit against Herman Sielcken and the whole syndicate. Chantland had decided that "anything which seeks to and does fix prices artificially at a different level is frowned upon and made illegal in most states by statute, and illegal by Federal statute." The wisdom of this move was questioned by the Secretary of State Philander C. Knox, who along with the president had not been informed of the move. He felt Wickersham had acted hastily and that Brazil likely had the right to valorize its coffee. The *Wall Street Journal* and the *Nation* also entered the fray and stated bluntly that the United States was getting payback for their previous, similar treatment of the Brazilians.

Much governmental and diplomatic wrangling ensued, with little effect. Norris introduced a bill to force the sale of surplus coffee, which went nowhere. Sielcken was brought before a congressional subcommittee investigating the Brazilian valorization scheme and brazenly denied that the syndicate was keeping surplus coffee off the market. The lawsuit filed by the attorney general dragged on for over a year, but was dropped by the new Wilson administration in 1913. In the end, Sielcken and his syndicate remained unscathed and sold the surplus coffee on their own terms. They became very wealthy through the valorization scheme, and the Brazilian economy was successfully propped up.

Colombian coffee hits the big time

While Colombia had vast volcanic landscapes highly suitable for coffee cultivation, it really wasn't until after World War I that coffee was grown in great quantities in that country (Beyer 1949; Pendergrast 2010). Two major obstacles had stood in the way of widespread cultivation: (1) the suitable coffee regions were almost inaccessible from the coast, requiring travel on the backs of mules across rugged, dense terrain or in boats down the wild whitewater of the Magdalena River, and (2) the country was almost constantly embroiled in civil wars with active fighting in 1854, 1859–1861, 1876–1877, 1885, 1895, and 1899–1903. The country was simply too exhausted and impoverished to develop a coffee industry.

When relative peace finally set in across Colombia in the early 1900s and world coffee prices hit record highs, the rich and poor alike began in earnest to plant coffee. Huge *haciendas* of more than 20,000 trees cropped up along the upper Magdalena River, which were broken into small tenement farms, where the ragged poor barely kept their heads above water. Numerous tiny, small-scale farms of less than a hectare spread across the mountains to the south. These small-scale farms came to dominate Colombian coffee production, with the farmers banding together to help harvest each other's fields.

New railroads were built and the Panama Canal opened in 1914, but most of the coffee was moved by mule to market from the small-scale farms. Colombia's coffee production rose steadily from about 200,000 bags in 1899, to 500,000 bags in 1905 and one million bags in 1915. Coffee exports came to represent 50% to 70% of all exports from Colombia. Brazil, desperately trying to limit its own overproduction to keep prices high, lost a quarter of its US market between 1914 and 1919 to Colombia and other countries that did not restrict their own production. Colombian coffee, grown mostly under forest shade, gained a reputation for its fine, flavorful quality and became highly desirable for upscale blends.

Brutal dictators with US support

Along with banana, coffee has held fast at the center of politics in Central America from the late eighteenth century until today. The two crops have dominated the export economy of the region and their growers have supported a chain of brutal dictators who have ruthlessly subjugated the native peoples (Wild 2004; Pendergrast 2010). These Indians have repeatedly revolted with bloody consequences. More often than not, Central American dictators have either been backed or covertly overthrown by the US government, under the guise of keeping communism out of the Americas. The history of the coffee republics is overflowing with examples of intolerance to opposition and bloody reprisal. Many of these instances were related in Chapter 2 on banana.

Guatemalan President Jorge Ubico eliminated all political rivals in the 1930s and early 1940s, by intimidation and assassination. He had at least a hundred student, labor and political leaders shot in 1933 when they had the audacity to strike and oppose his regime. He required that indigenous farmers work on coffee and banana plantations, and made his infamous Decree 1816, that made it legal to murder any Indians who did not comply. When Jacobo Arbenz gained power in Guatemala in the early 1950s and began reforms, he was overthrown by a CIA-engineered coup.

The United States worked hard for the next thirty years to support any and all Guatemalan dictators who were willing to back the policies of the United States and root out "communism." To support this effort, the US School of the Americas was established at Fort Benning, Georgia, to train Latin American soldiers in counterinsurgency techniques, including assassination, psychological warfare and interrogation tactics. Tens of thousands of trainees graduated from this school and returned to Guatemala and other Central American countries to support local dictatorships and fight insurgency.

In El Salvador in 1932, a band of Indians in the western highlands, led by the communist agitator Agustín Farabundo Martí, rioted and killed 100 owners and overseers with clubs and machetes. The response of President Maximiliano Hernández Martínez was to order the Civil Guard to murder over 30,000 Indians indiscriminately, some being tied by their thumbs before firing squads and others being shot above mass graves they had dug themselves. A particularly bloody, short-lived group of dictators ruled El Salvador after Martínez, using exported US tactics of torture and assassination. Perhaps the bloodiest was President José Efraín Ríos Montt, who in the 1980s ordered the deaths of thousands of left-wing agitators and Mayans living on the best coffee sites in the highlands.

In the 1930s in Nicaragua, President Anastasio Somoza Garcia's National Guard, trained and equipped by the United States, smashed all dissent by massacring scores of opponents and suspected rebels. The Somoza family held onto power for forty-nine years until they were finally overthrown by the Sandinistas, a party at least on the surface more supportive of the masses. The Sandinistas proved too liberal for the US military, which organized and funded a group in El Salvador called the Contras to overthrow them. The Contras were a nasty terrorist group who violated human rights with impunity, including assassination, rape, theft and the burning of schools and homes. When the US Congress finally got full wind of the Contras' atrocities and cut off support in 1985, the Reagan administration used illicit drug money to continue supporting the group. A peace was eventually brokered with the help of five Central American presidents, but a war-weary country voted the Sandinistas out of power in 1990 for a more conservative government.

The roller coaster of coffee prices

As the 1900s progressed, Brazil struggled long and hard to keep coffee prices high, mostly by itself (Bilder 1963; Bates 1998; Pendergrast 2010). Coffee production continued to rise at a rate faster than consumption. Brazil remained the dominant world producer, but numerous other significant players entered the market including Colombia, Guatemala, El Salvador, Kenya and Uganda. It was clear to the Brazilian producers that they had to continue to limit their coffee sales to keep prices high in this increasingly glutted world market.

Brazil revisited its valorization scheme, borrowing over £10 million in 1921 to buy and warehouse its overproduction. Brazilian president Artur da Silva Bernardes had huge warehouses built in 1922 to hold the vast quantities of coffee now being stored and required the planters to sell just enough coffee to meet demand. These tactics worked and the price more than doubled in the mid-1920s, in spite of increased production by other countries.

Wholesale grocers, traders, speculators, and roasters all howled, but the coffee planters grew very rich.

Unfortunately, the Brazilian growers grew overconfident with the heady feeling that this prosperity would never end. By 1928, over 100 million new coffee trees were planted, even though the warehouse now bulged with overproduction. The forest was destroyed at a rate of 3,000 square kilometers a year to make way for new plantations. Then, the unthinkable happened: the formerly friendly British and American bankers could see that the bubble was about to burst, and refused to loan any more money to buy the surplus. The Brazilian coffee market crashed on October 1929, a few weeks before the US stock market also crashed, leading to the Great Depression.

Stocks of coffee in Brazilian warehouses now stood at twenty-six million bags, about a million more than annual world demand. Coffee prices that were 22.5 cents a pound fell to 8 cents in 1930. Getúlio Vargas, who had overthrown the Brazilian government, banned new plantings and began burning the massive surplus, destroying about seven million pounds in 1930. Still coffee prices remained weak, as unrestricted Colombia and Central America sold more and more. By 1937, Brazil had destroyed tens of millions of pounds of coffee, but had lost about 15% of its US market to Colombia alone.

In response to pleas from Brazil, the other Latin American countries finally agreed to consider a quota system. The pressure from the African producers had become too great to ignore. A conference was held in Havana, but little was accomplished except a limit on the most inferior grades of coffee and a five-cent export tax on each bag to be used for advertising in the United States. In desperation, Vargas decided to open the floodgates and dump the Brazilian coffee onto the market.

Hitler soon after took Europe to war and the annual ten million bag market there dried up. Colombia, already severely hurt by Brazil's open door policy, asked the US State Department to help broker a quota system. Brazil's old idea did not look so bad anymore. The United States was ready to help, fearing the growing threat of communism in Central America and German sympathizers in Guatemala. A Pan American Conference met on June 10, 1940, and arduously hammered out an agreement where exports to the United States would be limited to 15.4 million bags, with Brazil allocated 60% of that total and Colombia 20%. Prices held tenaciously at an all-time low of 5.75 cents a pound in 1940, but finally doubled the next year.

In spite of these quotas, it became increasingly difficult to ship coffee to the United States, as the merchant marine was requisitioned during the war to deliver more critical foodstuffs, materials and troops. Soon coffee was being rationed to American households. When the Japanese bombed Pearl Harbor, Leon Henderson, the czar of the Office of Price Administration, froze coffee prices at 13.38 cents per pound to head off inflation, where it stayed until October 1946. Of course, the coffee producers suffered and they stopped expansion.

At the end of the war, the Brazilians were no longer producing a surplus of coffee. The fertility of their fields in Sao Paulo was exhausted and they were suffering from widespread infestations of the coffee bug, broca. As a result, prices rose to fifty cents a pound in 1947. In 1948, the United States allowed the Inter-American Coffee agreement to expire, but prices held reasonably strong as Brazil suffered a major frost and worldwide production continued to lag behind demand.

As planter confidence again surged, there was another worldwide planting spree, which ended predictably in 1954 when prices began to dramatically drop. African production was now in full force, tipping the scales toward another coffee glut. From 1954 to 1962, the price dropped 50%, from seventy-nine cents to thirty-five cents. A UN Coffee Conference was organized in July 1962 of the major importing and exporting countries to try to stabilize prices again. After almost around-the-clock negotiations, a tentative quota agreement was reached, giving the major producers in Brazil an annual quota of eighteen million bags and Colombia six million. A target price was set, which was maintained by quarterly quota adjustments that had to be ratified by two-thirds of the participating countries. This agreement survived many political upheavals and kept prices reasonably stable until 1989, when the United States finally pulled out. With the collapse of the Berlin Wall, they were no longer so afraid of the spread of communism in the Latin American states.

Change in the coffee landscape of northern Latin America

Traditionally, coffee was grown under shade, as part of the forest ecosystem. In the 1950s, attempts were made to convince growers in Latin America to "modernize" and grow their coffee in rows under full sun to maximize yields (Perfecto *et al.* 1996; Rice 1999; Wild 2004). This advice was largely ignored until the arrival of the leaf rust in Brazil in 1970 and Central America in 1976. The sad history of what had happened in southeast Asia a century and a half before energized many planters to change. It was thought that sun-grown coffee would stay drier than shade-grown coffee and thus would allow less germination of the pathogen. Also, chemical sprays could be more readily deployed onto plants well pruned and planted in rows.

By the end of the twentieth century, about 26% of the three million hectares of coffee in northern Latin America had been transformed into the "modern" system. The conversion rates ranged from 70% in Colombia to 40% in Honduras, 35% in Costa Rica, 17% in Mexico and 8% in El Salvador. The United States Agency for International Development (USAID) poured almost $81 million into this transforming effort, affecting half a million producers in the region.

Sun-grown coffee has proven to be more productive than that grown under shade, but at a higher cost. In comparisons of traditional, semi-modern (new varieties and some agrochemicals) and "modern" grown (sun-grown) coffee, average yields were 317, 953, and 1,397 kg/ha. The cost of producing a hectare of coffee in each system was $269, $1,092 and $1,739. Sun-grown coffee had higher planting costs and many more chemical inputs, and much more labor was needed to maintain the plantations.

In recent years, considerable concern has grown about the environmental impacts of sun versus shade grown coffee. The traditional approach is a passive production system with little chemical inputs, while modern approaches spend about 25% of their inputs on chemical sprays. The most dramatic impact of shifting from shade-grown to sun-grown coffee is on biodiversity. Most of the native forest trees are maintained in shade-grown coffee, while the forests are clear-cut to plant sun-grown coffee. As a result, shade coffee plantations are composed of nearly 60% to 70% of the total species diversity of native forests, while sun-grown coffee supports little of the original biodiversity. The growers of shade-grown coffee keep a rich array of tree species, which they use for fruit, firewood and construction materials. This arboreal diversity supports high insect species densities, which can be higher than tropical forests. Shade coffee farms also support most of the resident bird species found in the closed canopy forest, and migratory species use them seasonally at densities often higher than the native forests.

Shade coffee plantations provide a number of additional ecological services (see Perfecto and Vandermeer 2015). They are buffered against damage from high winds, hail and heavy rainfall. Shade reduces daily variability in humidity and temperature: midday temperatures can be reduced by as much as 5°C. The leaf litter under the canopy preserves soil moisture and is an important source of nutrients.

The local economies are also greatly impacted by the shift from shade-grown to sun-grown coffee. Traditionally, most coffee was grown by *campesinos* on small farms consisting of a single hectare or less. The bulk of the work on the farm was done by the campesino and his family and the profit obtained was theirs alone. With sun-grown coffee, huge numbers of laborers are imported to maintain and harvest the crop on vast plantations. The local economy shifts dramatically from family-based operations to corporate-run approaches.

The last thirty years have seen a strong niche market develop for specialty coffees that are based on social and environmental concerns. The first of these initiatives were the "fair trade coffees," which sought to provide farmers and agricultural workers in Latin America with better, stable prices and other resources to improve their lives. More recently, organic coffee grown on pesticide-free small farms has developed a strong niche market. "Bird-friendly" coffee brands have also emerged, touting the benefits of shade-grown coffee on migratory bird species.

Coffee today

For probably a century, coffee has been the most important tropical commodity sold in the world. It represents about half the trade of all tropical products (Anonymous 2003, 2015). The EU imports almost 70% of the coffee produced in the world, followed by the United States (24%) and Japan (7%). Per capita coffee consumption in the EU and United States has gradually declined since World War II, but worldwide consumption has grown modestly. Mass-marketed coffees in the United States have suffered most of the decline, while specialty coffees have grown in popularity. Overall, people in the United States drink about half as much coffee as in 1945.

The collapse of the International Coffee Agreement in 1989 opened the door to a new competitor: Vietnam. The country went from having a little over 1% of the world coffee market to over 12% in ten years, vaulting it to second place over Colombia as a world coffee exporter. Almost all the coffee grown in Vietnam is Robusta and 90% is cultivated by small farmers with 1–2 hectares; over four million people now depend on coffee for their livelihood there. The huge growth of the Vietnam coffee industry has had a significant effect on coffee price volatility by swelling oversupplies in years without an environmental calamity in the leading producer, Brazil.

Today, the Latin American coffee industry is in the midst of a huge battle with coffee rust on the vast acreages of Arabica grown (Arneson 2000; Cressey 2013; Malkinmay 2014). Over the last few years the disease has likely caused $1 billion in damage and has affected the livelihoods of 500,000 people, mostly small farmers and seasonal workers. Production in many regions has dropped by 10% or more annually, dramatically raising the market price for coffee. There have been other major rust epidemics in the region in the past, but this one is especially worrisome as plants in the higher elevation sites, which were largely spared before, are now being infested. This may be due to the rising temperatures associated with global warming.

A very promising resistant variety called "Castillo" has been developed in Colombia that could save that industry and perhaps others if it proves adapted to other countries. A major push is also now being undertaken at Texas A&M University to develop other resistant varieties that is supported by USAID, although it could take decades to develop new resistant varieties. The major source of resistance comes from the Robusta type, whose bean has a much poorer flavor than Arabica.

The treatment of coffee workers on large plantations varies widely, but most are still paid trifling wages and toil under harsh, unhealthy working conditions. Many bring their young children to help them in the fields to ensure that they meet their daily quota. Few have signed contracts to assure them of receiving their fair wages. A high proportion live in one-room dormitories with little privacy and only limited access to latrines and showers.

The long-term future of the coffee industry will likely continue to be a boom and bust situation. Environmental variation and political instability are not likely to go away in the future, and climate change may even lead to shifts in where coffee can be successfully grown. In spite of these difficulties, it is likely that a ready market of coffee drinkers will remain clamoring for the bean, encouraging coffee planters to continue taking the risk and growing the crop.

References

Adams, G.R. (1991) *Blacks and whites in São Paulo Brazil 1880–1980*, University of Wisconsin Press, Madison.

Allen, S.L. (1999) *The Devil's cup: A history of the world according to coffee*, Ballantine Books, New York.

Anonymous (2003) *Medium-term prospects for agricultural commodities*, FAO Commodities and Trade Technical Paper.

Anonymous (2015) *Coffee: World markets and trade*, Foreign Agricultural Service/USDA, Office of Global Analysis.

Anthony, F., Combes, M.C., Astorga, C., Bertrand, B., Graziosi, G. and Lashermes, P. (2002) The origin of cultivated *Coffea arabica* L. varieties revealed by AFLP and SSR markers. *Theoretical and Applied Genetics*, vol. 104, 894–900.

Arneson, P.A. (2000) *Coffee rust. The Plant Health Instructor*, doi:10.1094/PHI-I-2000-0718-02 Updated 2011.

Bates, R. (1998) Chapter 5. The International Coffee Organization. In: *Analytic narratives* (pp. 194–230). Bates, R., Greif, A., Levi, M., Rosenthal, J.-L. and Weingast, B.R. (eds.). Princeton University Press, Princeton, NJ.

Beyer, R.C. (1949) The marketing history of Colombian coffee. *Agricultural History*, vol. 23, 279–285.

Bilder, R.B. (1963) The international coffee agreement – A case-history in negotiation. *Law and Contemporary Problems*, vol. 28, 328–391.

Coppin, C. (1990) John Arbuckle: Entrepreneur, Trust Buster, Humanitarian, *The Freeman*, http://fee.org/freeman/john-arbuckle-entrepreneur-trust-buster-humanitarian/

Cressey, D. (2013) Nature news: Researchers marshal technology in bid to thwart fungal outbreak in Central America. *Nature*, vol. 493, 587.

Duncan, J.S. (2002) Embodying colonialism? Domination and resistance in nineteenth-century Ceylonese coffee plantations. *Journal of Historical Geography*, vol. 28, 317–338.

Hutchinson, L. (1909) Coffee valorization in Brazil. *Quarterly Review of Economics*, vol. 23, 528–535.

Lovell, W.G. (1988) Surviving conquest: The Maya of Guatemala in historical perspective. *Latin American Research Review*, vol. 23, 25–57.

Malkinmay, E. (2014) A coffee withers. Fungus cripple's coffee production across Central America, *New York Times*, May 5, 2014. https://www.nytimes.com/2014/05/06/business/international/fungus-cripples-coffee-production-across-central-america.html?_r=0

Pendergrast, M. (2010) *Uncommon grounds: The history of coffee and how it transformed the world*, Basic Books, New York.

Perfecto, I., Rice, R.A, Greenburg, R. and van der Voort, M.E. (1996) Shade coffee: A disappearing refuge for biodiversity. *BioScience*, vol. 46, 598–608.
Perfecto, I. and Vandermeer, J. (2015) *Coffee agroecology: A new approach to understanding agricultural biodiversity, ecosystem services and sustainable development*, Routledge, London and New York.
Rice, R.A. (1999) A place unbecoming: The coffee farm of Northern Latin America. *Geographical Review*, vol. 89, 554–579.
Sensabaugh, L.F. (1946) The coffee-trust question in United States-Brazilian relations: 1912–1913. *Hispanic American History Review*, vol. 26, 480–496.
Standage, T. (2005) *A history of the world in six glasses*, Bloomsbury, New York.
Thurber, F.B. (1881) *Coffee: From plantation to cup*, American Grocer, New York.
Ukers, W.H. (1922) *All about coffee*, Tea and Coffee Trade Journal, New York.
Wild, A. (2004) *Coffee: A dark history*, W. W. Norton, New York.
Wrigley, G. (1995) Coffee: *Coffea* ssp. (Rubiaceae). In: *Evolution of crop plants*, Second Edition. Smartt, J. and Simmonds, N. (eds.). Longman Scientific and Technical, Essex, England.

7 Rubber

Sources of rubber

There are hundreds of plants found across the world that produce a latex sap that can be made into rubber, but the species that has played the most important role in human history has been the Pará or hevea rubber tree (*Hevea brasiliensis*), a majestic tree extending to the top of the tropical forest canopy in South America (Priyadarshan 2011; Tully 2011). As will be told in the following pages, hevea was ruthlessly exploited in its native habitat in the late 1800s to early 1900s by rubber barons and their slaves, before being stolen from the wild to establish plantations all across the tropics. These plantations ultimately led to the collapse of the South American industry, but the brutal treatment of rubber workers has continued unabated.

There are a number of other latex producing plants that have been utilized by humans and will be described as our story unfolds. These include the castilla-rubber tree (*Castilla elastica*), found in southern Mexico, Central America and into Colombia; the gutta-percha tree (*Palaquium* sp.), which grows wild in Sumatra, Kalimantan and Malaya; the Ceará rubber tree (*Manihot glaziovii*) from Brazil; and the *Landolphia* vines found in Africa, Asia and Oceania. Added to this list during the desperate search for sources of rubber during World War II were guayule, and the Russian dandelion, called Kok-Saghyz.

Beginnings of rubber use

The production of rubber products was a skill known to the ancient Aztecs, Incas and Maya of Mesoamerica, Central and South America (Hosler *et al.* 1999; Tully 2011). The Mayans used latex from the castilla-rubber trees blended with morning glory sap to produce a wide array of products including arrow quills, shoes, religious objects, breastplates and waterproofed clothing. They were known to dip their feet into the milky sap and allow it to harden over smoky fires to produce waterproof shoes. It is unknown how they learned to combine the latex of the rubber tree with that of the morning glory, but the two plants live in proximity in the wild, and both were used in religious ceremonies, so it is not too surprising.

Rubber balls were by far the most popular ancient rubber product. Games were played with them by people all across the Americas and they were traded thousands of miles from their origin. Columbus was the first European to report on rubber on his second trip to Haiti. He observed the natives playing with rubber balls that had a far superior bounce to the European leather and wood balls of his crew. Columbus noted that the strange, elastic substance was produced from the milky latex dribbling from lacerated trees and bushes.

When Cortez invaded Mesoamerica, he discovered the Mayans playing what was probably the world's first team sport. The games were played with rubber balls in a long stone court shaped like a capital "I" with high, sloping walls (Whittington 2001). The courts varied greatly in size but averaged 120 by 30 feet, with rings suspended on each side. The goal was to keep the ball in the air, the players often using only their buttocks and knees, until someone got it to go through one of the rings or time expired. If no one got the ball through a hoop, the winner was the team who kept the ball most successfully off the ground and in the court.

Such games played an integral role in Mayan society. Thousands of such ball courts were built across Mesoamerica from about 1500 BC to AD 1521. The games were played in a raucous atmosphere, with numerous spectators and religious leaders in attendance, all betting on the outcome. The games were a very serious affair and could be a life and death struggle; the losing captain was sometimes sacrificed and his decapitated head kicked around.

Industrialization of rubber

While the Spanish and Portuguese explorers were the first to discover the use of rubber in the New World, they didn't see any great economic value to it, and it was left to the French to bring what they called *caoutchouc* to Europe's full attention (Loadman 2005; Jackson 2008; Tully 2011). Caoutchouc was probably derived from the native word, *cauchu*, meaning weeping tree. Charles Marie de la Condamine brought back samples of rubber to the Académie Royale des Sciences of France after his voyage in 1735 to Quito, Ecuador. He was sent to measure the length of a degree latitude at the equator to test Newton's theory that the earth was not a sphere, but bulged at the middle and flattened at the poles (Newton proved right!). In 1775, the French botanist Jean Baptiste Fusée Aublet was the first to use the genus name, *Hevea*, in describing a rubber tree found in French Guiana. The famous French chemist Antoine Francois de Fourcroy, who worked with Lavoisier to reform the system of chemical nomenclature, was among the first scientists to experiment with the properties of rubber in 1790.

The first widespread use of caoutchouc, or what became called rubber in Europe, was as an eraser to remove pencil marks from paper, much as it is used today. It is unclear who first used the term rubber, but Priestley, the discoverer of oxygen, widely popularized its use and called it India rubber because his sample had come from the West Indies.

Over the next few decades' other names for rubber emerged including gum-elastic, gutta-percha, Pará fine, seringa rubber and siphonia. India rubber and gum-elastic became a general term referring to all sources of rubber, while Pará fine, seringa rubber and siphonia represented rubber specifically from the Pará or hevea rubber tree of Amazonia. Gutta-percha came from taban trees found in maritime southeast Asia; it is a natural resin more like plastic than rubber and was used to make hard, resilient products rather than elastic ones.

In 1791, Samuel Peal discovered a way to attach rubber on a wide array of substances to make them waterproof and patented it. He dissolved the rubber in turpentine and then spread the sticky mass on leather, cloth, paper and even wood. It worked pretty well at waterproofing but was very smelly and tacky.

Thomas Hancock was the leading rubber inventor and entrepreneur from 1819 to his death in 1865 in England (Hancock 1857). He started out as coach maker and likely became interested in rubber as a way to keep his passengers dry. To reuse waste rubber, he developed a process by which maceration and heat was combined to produce a ball that could be molded into a plethora of shapes and sizes. He called his machinery a "pickling machine" to keep the process secret for many years without a patent.

In the early 1820s, Charles Macintosh began producing raincoats by dissolving rubber in naphtha and spreading the molten mess between layers of cloth. Macintosh and Hancock ultimately teamed up when it was discovered that Hancock's rubber was superior to that available to Macintosh. Unfortunately, the raincoats were smelly, and became hard and brittle in cold weather.

The big breakthrough in producing useful rubber products came in 1839, when a former hardware store owner in the United States, Charles Goodyear, discovered that when

Figure 7.1 Charles Goodyear, the inventor of vulcanized rubber

Source: US Library of Congress.

India rubber was combined with sulfur and high temperature, a much more resilient, longer lasting and much less smelly product was produced (McCain 2008). The process became known as vulcanization after Vulcan, the Roman god of fire. Goodyear had for five years fanatically researched by mostly trial and error, often in his kitchen, going deeply in debt and keeping his family in poverty.

In 1843 Goodyear finally took out a patent on the process, but never became rich from it, being so deep in debt that he could not put sufficient resources together to fully capitalize on his discovery. By then, Thomas Hancock also had learned how to vulcanize rubber, after receiving some samples of Goodyear's rubber. Hancock never called himself the inventor, but he conducted many experiments to duplicate Goodyear's breakthrough, and patented the process in England (a few weeks before Goodyear!). He made a fortune producing goods from it.

By the mid-1850s, the world could no longer run without rubber. An ever-growing array of products was being produced for the home and industry including hoses, shoe soles, sporting goods, hot-water bottles, gaskets, flooring, rubber bands, belting, insulation, balloons and even condoms. India rubber and then more successfully gutta-percha was used to insulate transcontinental telegraph lines. Every aspect of life was now touched by rubber products. The demand became even more insatiable with the invention of the automobile and the pneumatic tire at the turn of the century. Modern warfare fueled the demand as the rapid deployment of troops and equipment required the use of gas-powered vehicles built with myriad rubber products.

Wild rubber exploitation

To meet the ever-escalating demand for rubber, wild hevea trees were tapped across vast tracts of Amazonia in South America (Resor 1977; Loadman 2005; Hecht and Cockburn 2010; Tully 2011). Local farming was essentially abandoned as people went into the bush to collect rubber. During peak production in the region from 1890 to 1920, over 60% of the world's rubber came from an area about two-thirds the size of all of Europe. It encompassed parts of what are now Colombia, Ecuador, Peru, Bolivia, Brazil and Venezuela.

The rubber from the trees of the Amazon were called Pará fine, and was considered the highest quality rubber available. It derived its name from the Brazilian Province and city Pará (now Belém), which was the gateway to the vast Amazon valley. Oceangoing vessels could travel all the way from Pará to Iquitos on the Amazon River, a distance of almost 2,000 miles.

Rubber collection in Amazonia came to be dominated in the late 1880s and early 1900s by a handful of "rubber barons" who controlled millions of hectares and hired thousands of "rubber tappers." In Bolivia were Yaca Diaz and the Suarez brothers, Nicolas and Francisco. Colonel Resendo da

Silva ruled in Brazil, while Tomás Funes controlled much of Venezuela. The most ruthless of all was Julio César Arana in Colombia, called "Fat César." More on him later.

By 1912 there were almost 200,000 rubber workers in Amazonia, many coming from the northeast Brazilian province of Ceará, where severe droughts had left many desperate for food and work. They traveled long miles on river steamers loaded with livestock, many dying on route due to malnutrition and fever. They came as subcontractors that would be paid a percentage of what his boss or *patrão* obtained for the rubber, but in reality they became slaves. A tapper from Ceará would owe his sponsor the ticket price (and interest), the cost of everything he would need in his trade including the knives, cups and buckets to collect the sap, a rifle to defend himself and hunt, the rest of his food, a hammock and all the other humble furnishings of his wilderness home. Of course the tapper paid the patrão much more than he paid for it, often three to four times more.

The rubber tappers were called *seringueiros*, and they had a miserable life in the sweltering, insect-ridden, tropical lowland. They often lived totally isolated from family and all other humans in the dense forest among clouds of mosquitos, snakes, caimans and piranha. Festering insect bites, fevers and diarrhea were a routine part of the tapper's life, and an accident of any kind could prove fatal. Overall, the tappers lived a life of solitude, drudgery and indebtedness. Almost none ever returned home.

The seringueiros would be given a territory with a couple of winding paths through the forest where wild heveas could be found dispersed haphazardly, hundreds of meters apart. The tapper would hack out a clearing with a machete and build a primitive hut by felling trees and spreading a palm floor over stumps above the water line. A few smaller stems would be left standing to use as ridge posts and huge palm leaves would be laid across them for a roof.

The tapper generally had two long paths of around 150 hevea trees, which he would work on alternate days. He would cut slashes in the trees and place cups under them for latex collection. Often he would climb high up the tree using the previous cuts for hand and foot holds. When the slashing was done, he would return to each tree and pour the cups of freshly collected latex into a bucket that the tapper would lug through the forest from tree to tree. At the end of the day, he would build a fire and slowly cure the latex into rubber over the smoke. This process could not wait, as the latex is 60% water and spoils very quickly. He would pile palm nuts onto the fire to produce carbonic acid, which was needed to cure the rubber. The tapper would twirl sticks covered with the curing latex to gradually build up until he had fifty-pound balls of rubber. The whole process took about six hours and the tapper still had to prepare a meal and get at least some rest for the next day.

The work was seasonal, as the torrents of rain during the rainy season made collection impossible and the paths impassable. This was the time that the seringueiros would transport their rubber to market, floating the heavy balls down the many rivers of the Amazon basin. At the ports along the Amazon, the rubber would be graded for quality and then shipped throughout the world. The city of Manaus, located midway along the Amazon River, became the capital of rubber collection. It was transformed in the mid-1800s from a shabby, backwoods port to a major metropolis with paved streets, operas and fine restaurants.

Slavery in the Amazon

The vast quantities of rubber collected by the seringueiros was not enough to satisfy the greed of the rubber barons, particularly as world demand kept growing. As a result, they brutally enslaved the native peoples of Amazonia to solve their labor shortage (Loadman 2005; Jackson 2008; Tully 2011). The rubber barons did not do the work themselves, but hired men from the West Indies to do their dirty work. These men located and imprisoned the Indians and made them collect and process latex for rubber. The Indians were given a quota and then brutally punished if they fell short, often leading to their death. Wives were often held captive, and also tortured and raped, to keep their husbands working. If the slaves went missing, their whole families were captured and tortured to force them to return. Estimates vary, but it is quite possible that 250,000 people died in slavery in Amazonia, including tribes of the Witotos, Androque, Huitoto, Matses, Yagua, Karipuna and Boras.

Probably the most egregious of all the rubber barons was Julio Cesar Arana who controlled the Peruvian Amazon Rubber Company (PACO), encompassing a 12,000-square-mile tract of the forest. Along the Putumayo River at the then disputed border of Peru and Colombia, countless Huitotos were murdered or worked to death by Arana's people. For only meager rations and perhaps a gun, the Huitotos were forced to tap rubber in the dense forest for months at a time and lug huge loads of rubber for up to forty-five miles per day. If they did not meet their quotas or refused to work, they were savagely flogged and left to die, mutilated, castrated or sometimes burned alive. Some were even sold as slaves and used as unpaid prostitutes. When the rubber boom ended in Amazonia, less than 20% of the Huitotos were still alive.

Draconian exploitation and brutality also occurred during the late 1800s and early 1900s in the "Congo Free State" of King Leopold II, the second king of Belgium. There the local peoples were also forced to collect rubber from a number of native, latex-producing bushes and vines. Soon after he got control of the region, King Leopold decreed that all "vacant lands" would be considered state property and that the local people would be required to work up to forty hours per month on those lands to "support" the country. Of course, most of the profits went into Leopold's pocket. When this requirement didn't produce sufficient riches, the overseers gave the natives stiff rubber quotas that had to be met at the threat of life and limbs. This forced the natives to spend months collecting and smoking latex, ignoring their own farms and private industries.

The native labor was enforced by overseers and an armed colonial police called "Force Publique," Africans from tribes noted for their brutality. The overseers were paid bonuses based on the amount of rubber that was collected, which encouraged them to employ the same cruel techniques used in Amazonia to maximize their take, including beating, mutilation, kidnapping and rape. It is likely that more than half the population of the Upper Congo died under Leopold's rule.

A common practice was for the "police" to attack a settlement, take anything of value, put the women and children into a stockade and then give the male villagers a large rubber quota that they had to collect to ransom their families. Natives who rebelled or did not meet their quotas had their villages burned down, their children murdered, and their hands commonly cut off. Each policeman was issued a known number of bullets that were only to be used to kill unruly natives. No hunting for food was allowed. They accounted for the bullets they fired by removing and smoking the right hands of their victims and sometimes their penises.

The terrible practices occurring in the Congo and Brazil were ultimately brought to the world's attention through the tireless efforts of Sir Roger Casement of the British Consular Services. First as Consul to the Congo, he teamed up with the voracious newspaper editor Edmund Dene Morel to publish eyewitness reports and photographs of the horrors of the Congo. The resulting international outcry eventually convinced the Belgian parliament to remove Leopold from power in 1908. Later, as consul-general at Rio de Janeiro, Casement labored with the young American engineer/adventurer W. E. Hardenburg to get documentary evidence of the atrocities in Brazil published. Through their efforts, Fat Cesár was forced out of the rubber business by British courts in 1913, but he lived on as a rich, privileged man for decades. The center of rubber production had by this time shifted to the plantations of Asia, but the Brazilian atrocities continued for years.

Plantation rubber

Sooner or later, rubber was destined to become a plantation crop (Loadman 2005; Jackson 2008; Tully 2011). By the 1850s, the British had begun transplanting many other crops across their far-flung empire including coffee from Arabia to Ceylon, cocoa from Mesoamerica to Trinidad and sugar from southeast Asia to the Caribbean. Cinchona, the source of quinine water for the treatment of malaria, had also been spirited out of South America and established in large state holdings in India.

The first person to propose the planting of rubber trees into plantations was the Brazilian explorer and botanist João Martins da Silva Coutinho, who suggested the idea to the governor of Pará in the early 1860s. Brazil

Figure 7.2 Sir Henry Alexander Wickham, the man who smuggled rubber seeds out of Brazil

Source: US Library of Congress.

ultimately sent him to the Universal Exposition in Paris in 1867, where he served as chair of the rubber jury, which identified Brazilian hevea as the best source of rubber in the world. James Collins, curator of London's Museum of the Pharmaceutical Society read Coutinho's report and authored two papers himself recommending that rubber be grown in the empire's East Asia Plantations (like cinchonas!). By 1870, Clements Markham, head of the India Offices Geographical Department and secretary of the Royal Geographical Society, also became a strong proponent of plantation hevea and enlisted the help of Joseph Dalton Hooker, the director of Kew. Markham and Joseph Hooker's father, William Hooker, had played key roles in the heist of cinchona out of South America.

Markham and Hooker agreed that someone must collect seed in Brazil and ship it back to Kew for southeastern Asia distribution. Markham asked the British consul in Pará, Drummond-Hay, if he knew of anyone appropriate, and that person turned out to be the Englishman Henry Wickham, an adventurer and mostly failed Amazonian planter, who had experience tapping hevea. Drummond-Hay had been previously visited by Wickham, who had regaled him with stories of his journeys through Amazonia and offered to collect seed.

It took Wickham a couple of years to do the job, but in 1876 he delivered 70,000 hevea seeds to Hooker at Kew, after surpassing numerous obstacles. Wickham had to first identify a quality source for seed, which he found in the highlands above the Tapajos River some distance from Santarém. He had to be "Johnny-on-the-spot," ready to collect the masses of seeds raining down from the canopy, as they exploded from their seed cases. He had to keep the seed dry, as hevea seed rapidly germinate and spoil when moistened. Wickham did this by wrapping the dried seeds in banana leaves, packing them in baskets, and then hanging the baskets from the rafters of his cottage. In addition, he had the good fortune that little rain fell while he was collecting.

His final challenge was to get the seed to a navigable port and off to England. Here he was perhaps the most fortunate, if his account can be believed. There was a mostly empty steamship, *Amazonas*, at Santarém when his seeds were ready for transport. He asked the captain to take the shipment on credit, which was accepted, as the captain had already been swindled out of a cargo. Wickham and the British consul then had to convince the port authorities that they had loaded only "delicate botanical specimens" for Kew that were not subject to search. It was not technically illegal to transport hevea seed out of Brazil, but one would think that a sharp customs agent would have, at the very least, delayed shipment through red tape. When these last hurdles were cleared, Wickham and his wife hustled home to England with the seed, without so much as a word to their relatives who had accompanied them to the Amazon. They would all eventually scatter all over the globe, never to interact again.

While Wickham has gotten the lion's share of credit for spiriting hevea out of Amazonia, he was not the only person to import seed to England. Markham himself had shipped 2,000 seeds to Kew in 1873, but only a few of them germinated. Cross went to Brazil about the time as Wickham was returning and sent back thousands of viable seeds, whose seedlings wound up in botanical gardens all across southeast Asia. Charles Ferris, after corresponding with Cross, also shipped 2,000 seeds to Kew and some of these seedlings may actually have been the first hevea to arrive in Ceylon. Wickham's notoriety was based largely on the scale of his accomplishment and his own decades of self-promotion; he was knighted in 1920.

In spite of the heroic efforts of Wickham and others, it took almost thirty years for the first plantations of hevea to be developed. No one was sure how to go about it. Wickham was not given the opportunity to supervise the establishment of plantations, as Hooker did not consider him to be a sufficient horticultural scholar. Cross had the "proper" credentials and was taken seriously by Hooker and others, but he recommended the planting of hevea on poor, wet sites similar to those where he had collected seed in Brazil. This was a mistake, as hevea is actually much better adapted to the high, dry plateaus where Wickham had gotten his seeds. Hevea is also extremely slow to grow, taking 6–7 years before it can be tapped, which made the growers very reluctant to replace their already successful plantation crops with this unknown.

It took a devastating disease of coffee and a crash in the value of commodity crops in general to stimulate the planting of hevea. In Ceylon in 1869, as was previously described, the vast coffee plantations began to be attacked by a leaf blight and a decade later the industry was on its last legs. Planters tried to turn to cinchona, sugar, and tobacco, but prices were just not sufficient to make them economically feasible. They even experimented with hevea's faster growing cousin, Ceará rubber, but it proved ill-adapted and did not produce economic yields. Hundreds of planters went bankrupt and were forced to return to England. Those that ultimately succeeded planted rubber, or as we already discussed, tea.

In Malaya, hevea also had a strong advocate, Henry Ridley (or "Rubber Ridley"), the director of the Singapore Botanical Garden. Ridley took thousands of seeds from descendants of Wickham's plants in Ceylon, and painstakingly experimented for twenty-five years on its cultivation. He waxed on endlessly to anyone who would listen. Finally, in the 1890s, with the loss of coffee through disease and a collapse in tobacco, tea and sugarcane prices, the planting of hevea started to catch on.

The first significant acreages of hevea were established at the turn of the century in Ceylon and the Malay Peninsula, and by 1912 there were over a million acres of it spread across Ceylon, Malaya, Sumatra, the Dutch West Indies, Borneo, French Indochina and Australian New Guinea. The first plantation rubber began to appear in the world market about 1905, and by 1914 production from the plantations of southeast Asia exceeded that from Amazonia. The wild industry was effectively finished off in 1922, when its production fell below 10% of the world's rubber.

Big rubber companies enter the game

In the beginning, the vast majority of the rubber plantations were operated by British planters. In an attempt to stabilize prices after World War I, the British government instituted the Stevenson plan, which set up quotas to regulate production and thus maintain a high price. The reaction of the huge rubber manufacturing companies was predictably loud and negative, and they started to seek alternate sources of rubber in earnest (Tully 2011). Smaller native plantations began to flourish, and the big companies began to invest in their own plantations: Firestone in Liberia, Goodyear in the East Indies and Sumatra, Michelin into China and Cambodia, and Dunlop into Malaya and Sumatra. To a large extent, the big companies began to run local politics and set the standards for plantation worker treatment. Firestone ran particularly free in poor, debt-ridden Liberia, almost completely dictating government policy and dominating their export market.

Ford's big failure

A herculean effort to obtain alternate rubber sources was undertaken by Henry Ford (Jackson 2008; Tully 2011). He was convinced that he could tame the Amazon forest, by clear-cutting and replacing the native vegetation with hevea. He set out on a massive scale in 1928, building a city called Fordlandia with a hospital, cinemas, cafeterias, shops of all kinds, paved streets and tidy bungalows. He hired a workforce of 3,000 local peasants to live in his city, remove the forest, mill the wood and set the new plantations. Ford provided them with free medical and dental care, gyms, fair prices on commodities and a good wage.

Within a year almost 1,500 acres of rainforest were bulldozed and burned, and by 1934 another 9,000 acres had been removed. Tens of thousands of hevea seedlings were planted, some coming from seeds of ancestors of Wickham's trees in southeast Asia. Unfortunately, the whole operation was a dismal failure. The workers rioted over the Midwestern food and the rigid work schedule imposed upon them and at least two-thirds were fired or left. Most importantly, the rubber trees did poorly and the vast majority died. By clear-cutting the forest, a virtual desert was produced, with little nutrients available in the thin tropical soil and no protection from the tropical sun. Added to this, diseases and pests were able to rapidly hop from plant to plant, decimating huge surface areas rapidly. Ford finally gave up in 1945, after investing $10 million and never having visited the site himself.

The coolie labor force

As with the wild collection of rubber in Amazonia, the sad key to the profitability of plantation rubber was a ready source of cheap labor (Jackson 2008; Tully 2011). In the tropical areas where the plantations were being established, there were indigenous people available to work the plantations, but their lifestyle was closely tied to the ebb and flow of nature and the concept of hourly, regimented work was completely alien to them and even repugnant. This meant that the plantation owners either had to force them to work or search for labor elsewhere.

Several systems of labor recruitment emerged to meet the rubber plantation needs just like in the tea and coffee plantations. There was some day hiring of locals and others from away, but this was not considered optimum by the planters as the labor pool was too small and flexible to meet the constant demands of the plantation. What became much more popular was to hire contract or indentured workers from distant localities, where famine, overcrowding, poverty or just landlessness made people desperate for employment. The major recruitment points were first in China followed by India and to a more limited extent Java. India ultimately became the major recruitment point, with hundreds of thousands of people from the Tamil region finding their way to Ceylon, Sumatra and Malaya. In Australian Papua and New Guinea, the plantation owners were reluctant to import so many Chinese and Indians and instead legislated a tax on the locals, forcing them to work on the plantations because they had no other source of cash.

While the treatment of the workers in the plantations of southeast Asia did not reach the depravity of that in Amazonia and the Congo, verbal abuse was common as were beatings. The workers were considered second-class humans by the owners and managers of the plantations, totally disrespected and treated worse than pack animals. Deaths were common from disease and overwork, and it was not unusual to have an annual death rate at a plantation of 10% to 20% of the coolies or more.

Numerous laws were written all across southeast Asia to protect the workers, and many investigations were conducted, but the abuses long continued, with even the most inhumane perpetrators often receiving minimal punishments. Spontaneous riots and dissertations against the abuses became more and more common as the twentieth century progressed, but were generally met strongly by local police forces. It really wasn't until the middle of the twentieth century, after World War II, that the inhumane treatment of the workers was no longer commonplace.

German synthetic rubber

World War II turned the rubber industry upside down once again. As the war progressed on the eastern front, the vast plantations of east Asia were eventually seized by the Japanese and essentially laid fallow while the coolies were diverted into other jobs supporting the Japanese war efforts like the Burma Railroad (De Guzman 2008; Tully 2011; Solo 1954). This effectively removed two-thirds of the natural rubber out of the hands of the British and Americans. Stocks of rubber during the war years in the United States never exceeded a three- or four-month supply.

The Nazis never had a significant source of natural rubber during World War II, as their New Guinea and East African colonies had been stripped from them after World War I. The Germans did have a small synthetic rubber industry running by the end of World War I, largely through the work of Bayer chemists. They had three types of what was called "Buna" rubber based on different raw materials: H – a hard rubber used for casings; W – a soft rubber that could be used for tires but was inferior to natural rubber; and B – used for electrical insulation. German production of artificial rubber ceased after World War I, as it was more expensive and inferior to natural rubber.

Soon after Hitler obtained power, he realized that he would have to have synthetic rubber to wage war. He entered into an evil bargain with the German company I.G. Farben, which had absorbed Bayer and its rubber technology. The company contributed generously to the Nazi movement and many of its executives became Nazis themselves. Beginning in the mid-1930s, the company began to aggressively develop a series of synthetic rubber plants across Germany with government help, and in 1940 started work on its most monumental plant at Monowitz, in occupied Poland, near the infamous concentration camp at Auschwitz. This plant was nearly complete at war's end, at the cost of countless lives, but never produced any rubber.

The building of the rubber and gasoline plant at Monowitz signaled the most inhumane period of bondage associated with rubber production. In their greed to maximize profits, I.G. Farben had entered into a vile contract with Himmler's SS to provide slave labor to work at their plant. Several hundred thousand slaves worked at the Buna plants; some were POWs but the majority were Jews. The POWs were treated according to the rules of war, but the Jewish slaves were held at starvation rations until they died of malnutrition, disease, or simply could no longer work and were sent to the gas chambers. Many died within a few weeks, but the unlucky lingered on for three or four months until they were little more than emaciated skeletons. Most incredibly, they worked alongside other "free" foreign and German civilian workers, who rarely displayed even the smallest degree of kindness toward them.

The Germans also developed experimental farms near Auschwitz, where they used slaves to cultivate the rubber plant, Kok-Saghyz or the Russian dandelion. This effort was run by the agronomist Dr. Joachim Cäsar, a lieutenant colonel in the SS. This dandelion had been discovered in 1931 as a potential source of high quality rubber and scientists at the Kaiser Wilhelm Institute had been recruited to explore its real potential. Little rubber came from this effort, even though thousands of female slaves toiled and died in its service.

When the war finally ended, among those tried at Nuremberg were the twenty-four top executives of I. G. Farben. All denied they had anything to do with the atrocities conducted on the Jewish slaves, even though those slaves stood side by side with their other German workers, lived in prison camps within walking distance of the plant, and were sold to them for a fixed rate per head by the notorious SS. Only thirteen of these executives were convicted and none served more than a few years. Otto Ambrose, the director of the Auschwitz Buna plant, was sent to prison for only three years, and except for that brief period of incarceration, lived prosperously as an international consultant until the ripe old age of 89, never admitting or showing any remorse.

Synthetic rubber in the United States

The US solution to the rubber shortage during World War II was also to produce synthetic rubber, but it took them longer to get rolling than the Germans. They had a ready source of natural rubber from southeast Asia, so the stimulus was not high to develop a synthetic industry in the early decades of the twentieth century. An additional obstacle that arose during the war was the reluctance of Standard Oil to fully support that effort. Before the war they had obtained the rights to produce Buna rubber outside of Germany from I. G. Farben, but they were reluctant to support the synthetic rubber processes that did not require petroleum products. BF Goodrich and Goodyear were even challenged for patent infringements by Standard Oil in 1941. Standard Oil and I. G. Farben were tightly intertwined contractually and had worked closely together to control worldwide chemical markets. I. G. Farben was also the largest shareholder of Standard Oil, except for the Rockefellers.

The lion's share of US rubber was still from native sources in 1941, but by 1945 the vast majority of US rubber was synthetic. This turnaround came through US government encouragement and support of a consortium of rubber companies that partnered to produce commercial quantities of a general-purpose synthetic rubber, GR-S. This work was done in collaboration with a network of academic, government and industrial laboratories. Once a recipe for synthetic rubber was selected, production boomed when the government began paying to have plants built and then leasing them to the companies, who ultimately reaped huge profits. Over fifty factories were built between 1942 and 1945.

The United States also experimented with producing natural rubber from guayule, goldenrod and milkweed, but the synthetic effort far overshadowed it and these natural sources of rubber were ultimately abandoned. By the end of the war, the United States was able to surpass Germany's rubber production capacity and was actually producing more synthetic rubber than that from all the natural rubber plantations.

The rubber industry of today

At the end of the World War II, the natural rubber plantations of southeast Asia began producing huge quantities of rubber again, picking up where they had left off (De Guzman 2008; Tully 2011; Solo 2013). Even though vast quantities of synthetic rubber could now be produced, overall demand for rubber remained high, and for many applications, natural rubber is superior to synthetic rubber. Natural rubber is an extremely important component of tires and the majority of natural rubber now goes into tire production. Overall, natural rubber comprises 30% to 40% of the rubber used worldwide today.

Southeast Asia still dominates the production of natural rubber, generating 92% of the world's total. The major producing countries are now Thailand, Indonesia and Malaysia. Most rubber is produced on "smaller" farms, with over 75% of the plantations being smallholdings rather than estates. Rubber workers on estates still toil for minimal wages, with few benefits, long hours and inferior housing. While much clonal selection and some breeding has been done in the last few decades, almost all the rubber varieties now grown are derived from the seedlings collected by Henry Wickham.

References

De Guzman, D. (2008) History of the synthetic rubber industry, *ICIS Chemical Business Magazine*, Reed Business Information Ltd., United Kingdom.

Hancock, T. (1857) *Personal narrative of the origin and progress of the caoutchouc or India-rubber manufacture in England*, Longman, Brown, Green, Longmans and Roberts, London.

Hecht, S.B. and Cockburn, A. (2010) *The fate of the forest: Developers, destroyers, and defenders of the Amazon*, University of Chicago Press, Chicago.

Hosler, D., Burkett, S.L. and Tarkanian, M.J. (1999) Prehistoric polymers: Rubber processing in ancient Mesoamerica. *Science*, vol. 284, 1988–1991.

Jackson, J. (2008) *The thief at the end of the world: Rubber, power, and the seeds of empire*, Viking Penguin, New York.

Loadman, J. (2005) *Tears of the tree: The story of rubber – A modern marvel*, Oxford University Press, Oxford, UK.
McCain, D.R. (2008) Burning rubber, 1839. In: *It happened in Connecticut*, Morris Book, Guilford, CT.
Priyadarshan, P.M. (2011) *Biology of Hevea rubber*, CAB International, Wallingford, UK.
Resor, R.R. (1977) Rubber in Brazil: Dominance and collapse, 1876–1945. *Business History Review*, vol. 51, 341–366.
Solo, R. (1954) Research and development in the synthetic rubber industry. *Quarterly Review of Economics*, vol. 68, 61–82.
Tully, J.A. (2011) *The devils milk*, Monthly Review Press, New York.
Whittington, E.M. (2001) *The sport of life and death: The Mesoamerican ballgame*, Thames & Hudson, London.

8 Plantation crops

Yesterday and today

Ties that bind

All the plantation crops, while unique in many ways, share a surprisingly common history (Figure 8.1). First, clever people in antiquity recognized the potential value of each species and gradually improved them to serve humanity. The early sugarcane growers in New Guinea brought dramatic changes in the sugar content of sugarcane by starting new gardens year after year with only the sweetest individuals. Humans took the wild bananas in New Guinea with only a small amount of fleshy pulp and flinty seeds, and made them into an important starchy, seedless staple crop. Ancient farmers from the four corners of the earth gradually improved cotton by painstakingly selecting long staple types from wild species with only sparse amounts of short, coarse seed hairs. People in South America recognized improved hybrids of tobacco that occurred naturally and began cultivating them.

Almost all the species that became plantation crops were grown for thousands of years before European explorers discovered them and realized their potential economic worth. It wasn't until the early fifteenth century that the Portuguese began to build plantations of sugarcane on the Atlantic Islands along the African coast; large-scale sugarcane production soon spread from there to Brazil and the Caribbean. The Spanish introduced banana cultivation first to the Canary Islands and then the New World in the sixteenth century. When smoking took England by storm in the seventeenth century, a lucrative farming industry arose in the Virginia colonies. Coffee plantations were first established in southeast Asia by the Dutch in the 1700s, after a few seeds were spirited out of Ethiopia. When the British learned to make quality textiles in the mid-1700s, they shifted their source of raw cotton from India to plantations in the United States. China was the center of tea production until the 1800s when the British decided to establish their own plantations in India and southeast Asia. The first rubber plantations were planted in the 1850s by the Dutch and British in southeast Asia.

Once the large-scale commercial potential of plantation crops was recognized by industrialists, inventors took over and developed products that excited the masses. Numerous innovations made efficient cotton production

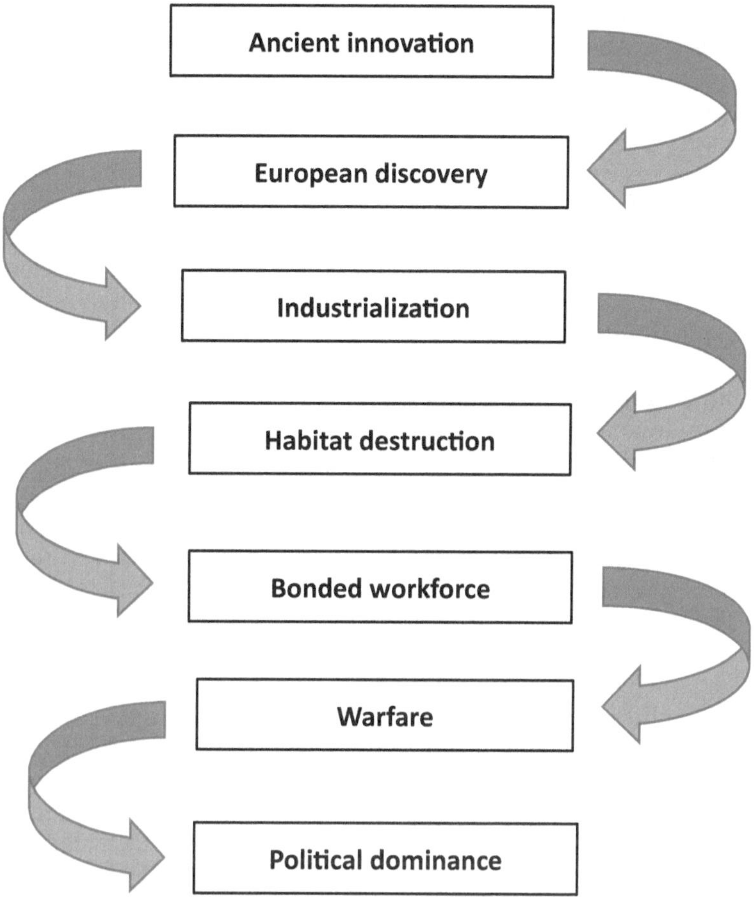

Figure 8.1 Path of development of plantation crops

possible in the mills of the Industrial Revolution including the flying shuttle, spinning jenny, water frame and power loom. The invention of the cotton gin made it possible to process the quantity of raw cotton needed to supply the needs of the textile factories. Coffee consumption was greatly boosted in the United States and elsewhere, when industrial coffee roasters and automated packaging machines were invented. Rubber became an integral part of the machinery of the Industrial Revolution, when the vulcanization process was discovered and produced a more resilient, longer lasting rubber that could take the wear and tear of pounding and pumping. To get bananas to their distant markets, entrepreneurs developed a network of railroads, refrigerated shipping fleets and warehouses to efficiently move the product.

As European and American demand for the plantation crops burgeoned, vast tracts of native vegetation were clear-cut in largely undeveloped areas of the world to establish plantations. Thousands of hectares of native vegetation were clear-cut to make way for banana in Central and South America. When the British demand for cotton burgeoned, the rich bottomland of the Mississippi Delta was converted to cotton in the years before the Civil War. The entire rainforest in the highlands of central Ceylon was completely denuded and planted with coffee and then tea. Thousands of hectares of rainforest were converted to rubber plantations in Malaya, Sumatra, the Dutch West Indies, Borneo, French Indochina and Australian New Guinea.

To maximize the productivity and profitability of the plantations, slaves or indentured servants were imported to maintain and harvest the crops. These peoples were brutally mistreated to keep profits high. Local people were initially forced to work the sugarcane fields and factories of the Europeans, but ultimately African slaves were imported to do the job. To harvest their tobacco, the southern colonists were the first to bring slavery to North America. When the United States became the major supplier of raw cotton to the burgeoning mills of England, they relied on relocated slave populations from the southeastern United States. To find enough laborers to work the British tea and coffee plantations in southeast Asia, the use of indentured Tamils from India became the norm. The Dutch paid local chiefs to force their subjects to harvest their coffee in Java. The awful task of clearing the forest and harvesting the banana plantations in Central America fell to desperate workers largely from the Caribbean, whose lives were completely controlled by the company. The Latin America coffee industry was largely built on the backs of the Mayans who were forced from their highland retreats to harvest the crops for a meager salary. When rubber plantations were established in southeast Asia by the British and Dutch, indentured workers were brought in from China and then India. In Australian Papua and New Guinea, the government legislated a tax on the locals, forcing them to work on the plantations.

Open warfare routinely broke out between the world's most powerful countries as they jockeyed for trade dominance. The high point of Caribbean sugar production coincided with a precipitous decline in Spain's power, and decades of almost continual warfare between France, England, Spain and the Netherlands. The warfare spilled over onto the islands of the Caribbean, culminating in a British blockade of sugar going to France. At the conclusion of the Seven Years' War, England gave up the whole of Canada for a few Sugar Islands. To restore their depleted coffers, the British passed the Sugar Act, which pushed infuriated colonists toward the Revolutionary War. The growing international demand for tea in the 1700s pulled a reluctant China into the worldwide trading arena. To get silver to pay for Chinese tea, the British began cultivating and selling opium from India, and when the Chinese emperor balked, the British started the Opium Wars. Tea plantations were established in India and southeast Asia in the 1800s, which soon eclipsed the Chinese trade, and contributed greatly to the political and social upheaval of that country.

The economies of the exporting countries became strongly tied to the profits of the plantations, leading to power struggles over who would control the massive wealth. In many cases, the developed countries importing the crop also shamelessly interfered with the internal politics of those undeveloped countries. A few banana industry leaders wielded great power by developing all-encompassing, vertically integrated companies and building extensive shipping empires. Countries in Central America became servants of the great banana companies, offering them sweet tax and land deals in return for bribes and infrastructure building. Coffee also held fast at the center of politics in Central America from the late eighteenth century until today. US-supported coups became commonplace whenever host countries "misbehaved" and began to cut into company profits. Wild fluctuations in the annual price of coffee led to Central American countries banding together to set quotas and the United States fighting them with international lawsuits.

The saga continues

Many plantation workers are still forced to labor in the fields and are trapped as virtual indentured servants after money is deducted from their substandard salaries for food, housing and other necessities. The rubber estates of Liberia have for a long time received international criticism for such abuses (Anonymous 2012, 2016). To build the country's economy, the Liberian government ordered local chiefs to draft laborers for work on the Firestone plantations in the 1920s. This system continued until the 1950s, and at its peak 80% of the workforce was in the fields through state-imposed coercion. Although this practice was stopped several decades ago, workers still remain bound by poverty and indebtedness. Compulsory deductions are taken from their paychecks for food, school fees and other necessities of life, and the tappers receive a penalty if they don't tap a required number of trees each day. To meet these quotas, workers often toil up to twelve hours a day, even though they are required to labor only eight hours.

Laborers on plantations all across the world are still woefully mistreated and suffer under oppressive, backbreaking and unhealthy conditions. Perhaps the most notable are the abuses being suffered by the tea workers in Assam (Chapter 4). These laborers have living and working conditions that are only marginally better than those in the mid-1800s. They live in broken-down hovels with no toilets and must defecate in the fields when nature calls. They are severely malnourished and many are ill without any health support. Children are forced to work along with their parents, and many of the laborers are required to spray dangerous pesticides with no protection.

Many plantation workers are hired as temporary workers from outside the host country, leaving them stranded and under the mercy of their employer.

A particularly startling example can be found in Guatemala, where a large number of the sugarcane workers come from Haiti (Anonymous 2016). These laborers are required to carry an identity card that ties them to a specific employer, forcing them to accept the abuses of their sponsor or be deported. Physical and verbal assaults are common, and some workers are actually kept under guard. In the age-old tradition of the plantation system, they become trapped in a cycle of indebtedness, paying for all their necessities at inflated prices. Their salary is tied to the amount of cane they harvest, requiring them to toil long hours under brutal work conditions to make even a small profit.

The use of migrant labor is not unique to the plantation system, as imported workers are the labor backbone of the fruit and vegetable industries in the United States and Europe. These seasonal workers are rarely subject to the same level of deprivation that occurs on the sugar plantations of Guatemala, but they do work long, backbreaking hours for very low wages and live in substandard housing. They also must suffer xenophobia and even racism from the locals.

The descendants of slaves and indentured servants brought in to harvest plantation crops have suffered persecution long after their families escaped the farms. Particularly notable have been the struggles of African Americans in the United States and the Tamils in Sri Lanka. In the United States, African Americans were long repressed after the Civil War by intimidation and laws. Their situation has greatly improved over the last half century, but major social stigmas still exist. In Sri Lanka, the native Sinhalese passed numerous oppressive laws keeping the Tamils as second-class citizens. In response, the Tamils fought a long and bloody war for independence in northeastern Sri Lanka where their numbers dominated. After decades of fighting, this conflict was only recently resolved.

Mechanization is rapidly eliminating the jobs of plantation workers, leaving scores without their traditional sources of income. Most dramatic have been the changes in the way sugarcane is harvested in Brazil and the rest of the world. Cutting, loading and transport of cane on many plantations is now done completely by machine. It has been estimated that in Brazil alone, 114,000 sugarcane cutters will lose their jobs between 2006 and 2020 (Scheidl *et al.* 2015). The shift to mechanization will open new higher tech jobs for machine operators and the factory workers building the harvest machines, but these jobs will come at the expense of hundreds of thousands of field labor jobs. Similar scenarios are being played out all across the agricultural sector, particularly in the fruit and vegetable industries.

In recent years, public initiatives have encouraged companies to develop supply chains that support small-scale farmers. These offer farmers a stable, fair price for their products and support environmentally friendly production practices. A strong niche market has developed in the United States and Europe for products from corporations showing such social and environmental responsibility (Wild 2004; Biswas-Tortajada and

Biswas 2015). The first initiatives to help small farmers started in Latin America with what came to be called "fair trade coffees." These provided small-scale farmers with stable, more equitable prices for their produce in an attempt to enhance their quality of life and support small-scale family farms. The environmental initiatives have focused on organic grown coffee and "bird-friendly" coffee brands. These coffees are produced on small-scale, pesticide-free farms under the forest canopy rather than in huge open monocultures.

Shifts in the supply chain from large to small farmers have not been restricted to coffee in Latin America. Small farmers are now responsible for most of the tea and coffee in Kenya, rubber in India and Indonesia, and sugarcane in Guyana (Herath and Weersink 2009). These initiatives demonstrate that smallholder farmers can support the large-scale production of export crops, although they remain a small percentage of the total.

Most of the plantation crops still remain vulnerable to devastation from diseases and pests, as only a limited number of varieties are grown. The banana plantations of Latin America continue to rely on "Cavendish," even though it is highly susceptible to the relentlessly spreading Panama disease. In China, India and Sri Lanka, vast acreages of tea are still planted with a single variety. Coffee rust threatens the major coffee varieties grown in Central and South America. Almost all the rubber varieties now grown in southeast Asia are derived from the seeds collected by Henry Wickham. A shift to greater crop diversity would have a long-term, positive effect on the sustainability of the plantation system and stimulate native plant and animal biodiversity, soil conservation and landscape aesthetics. (Perfecto and Vandermeer 2015).

Plantation crops continue to be at the center of international disputes. In particular, the water demand for cotton and other crops in Egypt and Sudan has sparked diplomatic warfare all across the Nile basin (Appelgren *et al.* 2000; Abbas 2016). A treaty was signed during British colonial rule that set aside 80% of the Nile's water for Egypt and Sudan. The seven countries that are upstream (Burundi, Ethiopia, Kenya, Tanzania, the Democratic Republic of Congo, Rwanda and Uganda), now demand greater access to that water. The upstream countries argue that the treaty is an unfair leftover of colonialism, while Egypt and Sudan contend that they rely on the Nile as their sole source of irrigation water and the other countries have plenty of other water resources. The building of the Renaissance Dam in Ethiopia has sparked intense, angry debate (Cambanis 2010; Hussein 2016).

The plantation crops still hold sway over the economies of many developing countries, giving the owners of the largest plantations considerable political power. Perhaps the days are gone when a company completely ruled the affairs of countries, but many growers and marketers still hold major political clout in Latin America and Southeast Asia. This power, and the desire to keep profitability high, has contributed to the slow pace of change in the conditions of plantation workers.

Déjà vu

The environmental and human abuses associated with the plantation system are being repeated all over again as world demand for palm oil has rocketed (Koh and Wilcove 2008). This vegetable oil is now found in probably half of the processed food and household products in United States, Canada, Australia and Europe. The amount of oil palm cultivation land increased from 3.6 million hectares in 1961 to 13.2 million hectares in 2006, and the growth of the palm oil industry continues unabated. Vast tracts of tropical rainforest are being ravaged to make way for oil palm plantations in the two largest palm oil–producing countries, Indonesia and Malaysia (Sheil *et al.* 2009). Over 50% of this expansion has come at the expense of forests. Bird and butterfly species diversity has dropped by 75% where this devastation has occurred, and orangutan and Sumatran tigers are on the verge of extinction.

As plantations systematically replace the rainforest, the local people who had relied on them have no choice but to work on the plantations. They toil under hot, degrading conditions for meager salaries that barely allow them to support their families. In many cases, their children join them in their backbreaking labors without pay. Local governments are doing little to combat this human and environmental exploitation, enjoying the graft and profits flowing from the oil palm industry.

The widespread growth of the plantation system is not restricted to oil palm. Stimulated by the exponential growth of the biofuel industry, large corporate entities are currently buying huge swaths of land in Guatemala, Malawi, Mozambique and elsewhere to establish sugarcane plantations (Chinsinga *et al.* 2013; Durr 2016). These so-called land grabs rely on government support to displace indigenous people and destroy the native habitat. These large-scale land acquisitions present short-term benefits to the local communities in the form of jobs and the capitol for rural development, but destroy local social systems and make them dependent on outsiders for their livelihood. The real financial gains pass to corporate entities outside the local community. This form of "neo-colonialism" is not limited to the plantation crops, as supermarkets in Europe and the United States also outsource a significant proportion of their fruits and vegetables from large corporate farms in developing nations.

And so history repeats itself. The expansion of the plantation system today is following the same script played out in the past, starting with sugarcane in the 1600s, banana, tobacco, cotton and coffee in the 1700s, and tea and rubber in the 1800s. As the popularity of each plantation crop exploded, human greed led to the widespread destruction of native habitats and the cruel enslavement of a workforce. Plantation crops turned social structures upside down leading to revolution and government change. Many men became vastly wealthy from them through graft, bribery and unthinkable cruelties. These sad practices will continue as long as profit trumps conscience and no one cries foul.

References

Abbas, A. (2016). Ethiopia close to finishing 70% of Grand Ethiopian Renaissance Dam, *Daily News*, Egypt. http://www.dailynewsegypt.com/2016/05/28/ethiopia-close-to-finishing-70-of-grand-ethiopian-renaissance-dam/

Anonymous (2012) *Rubber production in Liberia: An exploratory assessment of living and working conditions, with special attention to forced labor*, Verité, Martin P. Catherwood Library, Cornell School of Industrial and Labor Relations, http://www.verite.org.

Anonymous (2016) *Strengthening protections against trafficking in persons in federal and corporate supply chains: Research on risk in 43 commodities worldwide*, Verité, Martin P. Catherwood Library, Cornell School of Industrial and Labor Relations, http://www.verite.org.

Appelgren, B., Klohn, W. and Alam, U. (2000) *Water and agriculture in the Nile basin*, Food and Agriculture Organization of the United Nations, Land and Water Development Division, Rome.

Biswas-Tortajada, A. and Biswas, A.K. (2015) *Sustainability in coffee production: Creating shared value chains in Colombia*, Routledge, London.

Cambanis, T. (2010) Egypt and thirsty neighbors are at odds over the Nile, *New York Times*, September 26. http://www.nytimes.com/2010/09/26/world/middleeast/26nile.html.

Chinsinga, B., Chasukwa, M. and Zuka, S.P. (2013) The political economy of land grabs in Malawi: Investigating the contribution of Limphasa Sugar Corporation to rural development. *Journal of Agricultural and Environmental Ethic*, vol. 26, no. 6, 1065–1084.

Durr, J. (2016) Sugar-cane and oil palm expansion in Guatemala and its consequences for the regional economy. *Journal of Agrarian Change*, doi:10.1111/joac.12150.

Herath, D. and Weersink, A. (2009) From plantation to smallholder production: The role of policy in the reorganization of the Sri Lankan tea sector. *World Development*, vol. 17, 1759–1772.

Hussein, W. (2016) Egypt's water minister says country facing "critical phase," *Al-Monitor*, Egypt, April 10. http://www.al-monitor.com/pulse/originals/2016/04/egypt-water-minister-interview-nile-drought-ethiopia-sudan.html.

Koh, L.P. and Wilcove, D.S. (2008) Is oil palm agriculture really destroying tropical biodiversity? *Conservation Letters*, doi:10.1111/j.1755–263X.2008.00011.x.

Perfecto, I. and Vandermeer, J. (2015) *Coffee agroecology: A new approach to understanding agricultural biodiversity, ecosystem services and sustainable development*, Routledge, London.

Scheidl, H.A., Simon, A.T., Pacagnella Junior, A.C. and Salgado Júnior, A.P. (2015) Environmental impacts of mechanization in Brazil's sugar and ethanol industry: The cutting, loading, and transportation process case. *Environmental Progress & Sustainable Energy*, vol. 34, no. 6, 1748–1755.

Sheil, D., Casson, A., Meijaard, E., van Nordwijk, M., Gaskell, J., Sunderland-Groves, J., Wertz, K. and Kanninen, M. (2009) *The impacts and opportunities of oil palm in Southeast Asia: What do we know and what do we need to know?* Occasional paper no. 51, CIFOR, Bogor, Indonesia.

Wild, A. (2004) *Coffee: A dark history*, W. W. Norton, New York.

Index